ANCIENT INDIAN METALLURGY
(THEORY AND PRACTICE)

ANCIENT INDIAN METALLURGY

(THEORY AND PRACTICE)

ASHOKA K. MISHRA

AGAM KALA PRAKASHAN

NEW DELHI

First Published 2009

ISBN : 978-81-7320-092-2

Published by:
Dr. Agam Prasad, M.A., Ph.D., Diploma in Museology
For **Agam Kala Prakashan**
34, Central Market, Ashok Vihar
Delhi - 110 052
Tel.: 27212195, 47098 550, 65688806-7
Fax: 91-11-27212195
Email: agambooks@gmail.com, agambook@yahoo.com

Composed by:
Aman Printographics
Delhi
Mobile: 9891395309

Printed by:
Salasar Imaging System
Delhi-110 035

Printed in India

PREFACE

The culture of every nation derives strength and vitality from its traditions and achievements of the past. In metallurgy, we are here to a great tradition. Metallurgy is the study of metals including their extraction from ores and the modification of alloys to produce materials of superior properties like bronze, brass, steel, etc.

Archaeologists frequently study the technological system and products of past populations to understand the equipment, adaptation or economics of these groups, or to use the artifact complexes as ethnic markers or indicators of chronological horizons. In spite of the fact that millions of artifacts have been excavated, described, and analyzed, information about how they were produced is often lacking or limited to the more superficial features of the pieces. Materials analysis, the use of ethnographic and historical accounts of the production of similar artifacts, and the experimental replication of specific artifacts are all valid analytical approaches, but they often treat the artifact out of context. But their metallurgical analysis play a pivotal role for understanding the technology of the past.

The metals, most commonly used in archaeological contexts were copper alloys (such as bronze), silver, gold, tin, lead and zinc under non-ferrous group and iron steel under ferrous group in which Iron Age dawned very late in contrast to Copper Age due to its technical complexity of extraction, albeit the ratio of iron & copper ore in the earth's crust is 10:1.

Daniel Wilson (1816-1892) was the first British archaeologist who apply the Danish approach to prehistoric archaeology based on Christian Thomsen (1788-1865) three age system of classification (Stone, Bronze and Iron Ages) in his book *Archaeology and Prehistoric Annals of Scotland* (London, 1851) which divided into four sections assigned to the period of Stone (Primeval), Bronze (Archaic) and Iron (Tentonic) and to Christian era. While Wilson applied the Thomsen-Worsaae chronology of Danish Prehistory to Scottish data, there is no evidence that he clearly understood the seriational principles upon which Thomsen has based his chronology. After Sir John Lubbock, later Lord Avebury (1834-1913) who suggested the sub-division of Stone Age into an earlier Paleolithic and later Neolithic, the terms Chalcolithic, Encolithic and Copper Age are used which were interchangeably by some authorities to describe communities between the Neolithic and the Bronze Age which used unalloyed copper rather than true tin bronze. The chalcolithic (chalcopyrite ore of copper + stone) term is much less widely used than other divisions and subdivisions of three age system, partly because of difficulty in distinguishing copper from bronze without chemical analysis and partly because many areas did not have a chalcolithic period at all. Different wages have grown up in different areas and this can cause confusion: for instance, the Italian and Spanish Chalcolithic or Copper Age cultures are equivalent—both chronologically and technically to the Early Bronze Age in the Aegean.

The metallurgical analysis of archaeological materials appears to have started at the end of the last century in England with Professor William Gowland's pioneer work on upon some superficial examination of metals from archaeological excavation in 1920, when we find in voluminous excavation reports that metallic finds are well illustrated and described, but their composition is dismissed with three to four analyses and little comment upon the nature of the metals. The more extensive research in the field of archaeometallurgy may be said to have after the end of the First World War. In England, it was during the years between the two wars that the British Association Summerian Copper Committee was founded under the leadership of Professor C.H. Desch who carried out a long series of analyses of metal objects from many countries. In Germany, Witter and Otto published a great series of analysis of German material. After Second World War, splendid programme of metallurgical research carried out in Austria under the direction of Professor Pittioni and his colleague in Germany Dr. Junghans and Sangmeister are very actively engaged in research and are adding much new material. Dr. Andreas Oldenberg has also greatly added to our knowledge of prehistoric metallurgy in northern Europe, while Dr. Pleiner has been very actively engaged in the study of iron technology in Czechoslovakia. In Italy, Professor Mario Bertolone has founded a Scientific Study Center at Varese, where in collection with engineer Dr. Storti, a very considerable programme of metallurgical research is in hand. Again in Italy, outstanding analytical research is in progress by the Centro per la Storia della Metallurgia of the Associazone Italiana di Metallurgia di Milano. Attention is also given to the metallurgy of the Ancient East by Professor Barridwood's research upon Syrian copper and bronze, while Professor J. Needhan and Dr. Wang Ling have opened a new field in the metal technology of Ancient China. In England the University Research Laboratory for Archaeology and History of Art in Oxford and the Ancient Mining and Metallurgy Committee of the Royal Anthropological Institute, London; all with active work in progress.

When a metallurgical analysis is initiated, it is of the first importance to consider what elements in the metal should be sought for. In early days of more or less isolated research the chemist often looked for, or at least published, only those elements in which he was interested for his particular problem. In those days it was not realized to what degree research would develop or that other workers would very probably require to know of the presence or absence of further elements. This is a matter in which international collaboration would appear to be very necessary, since it is very discouraging to the general scholar when comparing analysis to come up against lack of uniformity. For example, an element in which he is interested may be quoted by one authority but omitted by another. Since trace elements (elements which on chemical analysis make up less than 0.01 per cent of a material) are of interest and importance in research upon ancient metals, it is important not only to seek for the usual elements covering the major constituents of prehistoric metals, but also to give adequate cover for those elements which may appear only as traces and which indeed may be of decidedly rare occurrence within the limits of our present knowledge.

Another matter which may be mentioned here, and which presents certain practical difficulties, is that of limits of detection of spectrographic work. When we are considering mere traces of an element present in metal, the accuracy of the percentage figure recorded will depend upon the possible limit of detection. This limit varies from different elements, but it also varies as between different spectrographers. Such variations cause difficulty when we endeavour to make close composition between trace elements quoted in analyses by different scholars. For instance, it is not always easy to know just what is meant by the words "trace" and "not detected". One sepectrographer may use the term "trace" for an element which is present in metal but in quality of say less than 0.01%, while for the same element another may quote a definite percentage by X-Ray diffraction method figure of a lower order. Again, the term "not detected" can be confusing. Clearly the element has not been observed by the spectrographer, but it might be detected by another operating with a more sensitive limit of detection. Of course, on the other hand, the element may be entirely absent from the metal. In this connection, a most helpful system is to place the limit of detection for the element concerned in brackets after not detected. As I see it, the more important aims for the metallurgical analysis of archaeological material are: (a) to obtain data from the various metals through which they may be related to mineral deposits, so establishing likely sources for the raw material. To this end we require research and analysis upon the relevant ores from the countries concerned. (b) To obtain information concerning the structure of manufacture of certain archaeological artifacts. This is an exceedingly important branch of the subject, and here metallographic examination of antiquity is essential because it can tell the archaeologist a great of many things of first importance concerning the history and development and ancient metal technology and the relationship between the extracted metal and ore.

In Indian context, among the earliest contributors (1920-1947) to the field of ancient metallurgy have been Mr. M. Sana Ullah, Drs. Hamid and Parmsivan of the Archaeological Survey of India who reported on the chemical composition of metals from Harappa (36°38'; 72°52'), Mohenjodaro (27°18'; 68°07') and Taxila(33°47'; 72°49') . Dr. B.B. Lal of the same department from 1947-1965 has done quite a good deal of work in this direction. Among Indian universities, work has been done by Professor Satya Prakash and his co-workers Drs. N.S. Rawat and Rajendra Singh on the chemical studies on ancient Indian coinage. Professor. K.T.M. Hegde of M.S. Univesity, Baroda, has studied copper metallurgy of Chalcolithic period of Gujarat and Rajasthan and made technical studies on iron, lead, silver and zinc metallurgy from 1965 to 1985. His main contribution in the field that he compared the composition of the various metals found in one or more countries with each other, so establishing groups of metals which relative agreement in composition is found and by which it may be possible to derive from similar sources. This research throwing a fresh light upon cultural as well as technological problems. Dr. D.P. Agrawal of Tata Institute of Fundamental Research, Bombay, for an assessment of the Harappan Copper-bronze technology and ecology is still the best one. He suggests a high probability of the extensive use of oxide ores even though the use of the sulphide ore was probably known right from the even beginning indicating the use of mature copper technology from its very inception. Among the copper-using cultures of inner India the most important recent find is that of Jodhpura—Ganeshwar culture in northeast Rajasthan by R.C. Agrawal. The early metallurgical significance of Rajasthan is not only limited to the Jodhpura-Ganeshwar complex to the northeast Rajasthan. H.D. Sankalia's escavation at Ahar (27°42'; 75°38') in southeast Rajasthan throw an effective knowledge of copper metallurgy in the whole of the old Mewar area between Udaipur and Bhilwara. K.T.M. Hegde's work in the *Journal of the Oriental Institute*, Baroda in 1964 and another work by the author in 1992 confirmed that Ahar could be a source of copper artifact and metal supply to the neighbouring chalcolithic culture of Malwa. J.C. Brooke's report on 'The Mines of Khetree in Rajputana' published in *Journal of Asiatic Society of Bengal* in 1984 is the most scientific report on ancient copper mining of Khetri which confirm its relation with Kolihan mine.

In the field of ferrous metallurgy, its metallographic analysis has remained sadly neglected in India, and only early studies are by O.P. Agrawal on the material from Atranjikhera (27°42'; 78°44'); H.C. Bharadwaj on Rajghat (25°4'; 83°1') and the megalithic material from Tadakanahalli (14°21'; 74°52') in Karnataka (both ferrite and pearlite) and A.K. Ghosh and P.K. Chattopadhyay's study on iron objects of Barudih in Jharkhand. Chemical and metallographic studies have been undertaken by V.D. Gogte on iron specimens from megalithic sites of Mahurjhari (21°14'; 79°30') and Naikund (21°15'; 79°31'). The metallographic studies on the iron specimens of historical period have been summarized by P.K. Chattopadhyay and Andrew Mc William's work on Delhi iron pillar. So far as the archaeological occurrence of iron and steel is concerned, a detailed review has been published by Professors B. Prakas and V. Tripathi of Banaras Hindu University in 1986 and Arun Kumar Biswas in *Minerals and Metals in Ancient India* in 1996. He tried to correlate the ancient Indian literary sources with archaeological evidences in the field of metallurgy. In the area of ethnoarchaeology Verrier Elwin in 1942 recorded a detailed study about the 'Agraria' tribe and their smelting process in India in the Agaria belt in Maharashtra, Madhya Pradesh and Jharkhand. M.K. Ghosh and K.N.P. Rao in 1964 of Tata Steel, Jamshedpur attempted to locate those iron craftsmen of Munda and Asur tribes near Jamshedpur and two places in Orissa, who could remember the production technology of ancient iron. These persons were successful in operating three different furnaces one each at Kamarjoda in Jharkhand and Jiragora and Chiglabecha in Orissa. This was followed by B. Prakash and K. Igakkhi in

1984 at loharpura in Bastar and by author in different localities in Chhotanagpur and Bastar region in 2003 in his work *Ethno-archaeometallurgy of Iron in India*. In 1990 M. Sharma surveyed eastern Madhya Pradesh and located too old men who remembered the ancient iron making process and after long persuasion they constructed a furnace and operated it successfully at Bishunpur under the project 'Vikas Bharti'. Later on a detailed survey of eastern Madhya Pradesh and neighbouring area of Bihar and Uttar Pradesh was done to locate 'Agarias' and their furnaces by A.K Mishra and V.K. Chaube from 1993 to 1996 of Institute of Gandhian Studies, Varanasi with collaboration of I.I.T., Mumbai. Recently National Metallurgical Laboratory, Jamshedpur got interested in the study of heritage not only in the area of ferrous and steel but also in the non-ferrous melting and casting area.

In writing this little work I highly beholden to Professor Ram Adhar Pandey who is currently my H.O.D. for his painstaking and cruel editing of the entire manuscript. Swetank diffused my scientific confusions related to this work and Mr. P.K. Chatterjee typed the contents of this work without errors. For publishing this exercise I am thoroughly grateful to Dr. Agam Prasad.

Ashoka K. Mishra

History, Culture and Archaeology
Dr. Ram Manohar Lohia Avadh University,
Faizabad

25 December, 2008

CONTENTS

CHAPTER 1

INTRODUCTION

Metallurgy is the science and technology of metals, means it is concerned with the production of metals from their ores, the purification of metals and the preparation of alloys to produce material with superior properties. Metallurgy depends upon the ability to smelt ores into metal. The term, however, is often broadened to include the working of metal using heat. We will employ the term in this broader sense.

Process metallurgy is concerned with the extraction and production of metals while physical metallurgy is concerned with the mechanical behaviour of metals. Here metal cannot be rigorously defined, because earlier attempts to define it in terms of ductility, metallic lustre, conductivity, etc., have failed due to anomalies or inability to be exclusive, and the chemical definition of a substance whose hydroxide is alkaline[1] fails since it does not define materials having commonly accepted 'metallic' properties. In general metal enters chemical reaction as positive ion or cations.[2] There are certain elements such as tellurium, which have the physical properties of a metal and chemical properties of a non-metal. The distinction between the two groups (metal and non-metal) is not sharp. The typical metallic structures are the body-centered cubic, the face-centered cubic (cubic close-packed) and the hexagonal close-packed structures, hep etc. A more acceptable approach is to consider that a metal is a crystalline material in which ions[3] are connected indirectly through the field of free electrons surrounding them. Each ion attracts as many neighbouring ions can, giving a closepacked structue of short bonds. Therefore, good strength and relatively high density , associated with good electrical and thermal conductivity, ductility and reflectivity.

Metals found in nature in the metallic state rather than as ores were used by primitive men long before the art of smelting was mastered. Gold and silver occur in nugget form, as does copper. Native gold and silver were used in many parts of the world where the metals were found, but they did not, in themselves, give any marked stamp to a culture. Native or "float" copper, on the other hand, was the basis for substantial developments in the upper Mississippi River valley tradition of North America. The Old Copper Culture made use of the metal, nonmetallurgically it would seem, to produce tools, weapons and ornaments. Gold and silver were used for little more than ornaments, as in Indus Valley and copper artifacts, however useful, were a bit soft, but premetallurgical iron was known and worked. This iron was rare, coming from meteorites (an unreliable source at best), but greatly appreciated. In fact, the meteoric metal was of far better quality, being a nickel steel, than men were able to smelt before recent historic times because it contains moderate amount of nickel and smaller amount of carbon.

Iron of this nonmetallurgical sort made a great deal of difference to the Eskimo of Greenland, Ellsmere, and Baffin Lands who exploited an iron meteorite of many tons' weight. It had been worked smooth, with all projections broken off, before it was moved in this century to the American Museum of Natural History in New York. Rarely does a museum capture in its entirety an aboriginal iron "mine".

These nonmetallurgical approaches to metals were based upon their treatment as malleable stones. The bits were hammered, broken, and ground into shape. Iron pieces were so small and rare that they were used only for crucial parts of artifacts, and small, sharp pieces were set into handles in the style of the microliths of Mesolithic times.

Whatever the technical limitations in handling these metals it was recognized that they were something extraordinary. Some became valued as precious metals. Utilitarian considerations gave copper and allied metals (tin, antimony, arsenic, zinc) their worth. Iron, possibly because of its magnetic variety of magnetite ore (the lodestone) and possibly because of its sometime extraterrestrial origin, was held in special esteem as having magical powers. In time a considerable folklore has built up around the metals.

Ancient metallurgy in the proper sense had two origins according to our present understanding. One of these, responsible for Old World developments in all metals, lay in the near Eastern-Anatolian area and India. The other, of which we are less sure, probably originated in Peru and gave rise to limited nonferrous metallurgy in the New World. While particular traits and achievements may mark developments in this region or that, ultimately the ideas are traceable back to these two places. There is at present no known prehistoric connection between the metallurgies of the two hemispheres i.e. Old and New World.

The dating of these events is more uncertain than their location. In part this uncertainty arises from the sparseness of evidence and the gradual acquisition of the art. In the Near East and India copper was in use by 7000 B.C.E. with some assurance of the use of heat a millennium later and smelting by 6000 B.C.E.. By 3500 B.C.E. metals had a substantial impact on the cultures of the Near East and India and were spreading outward in all directions. In the New World metallurgical developments started between 750 B.C.E. and the beginning of the Christian era. Some problems of dating are coupled, in this estimate, with little assurance that all the evidence has been gathered. A rapid northward spread toward Middle America, however, seems certain.

The limits of prehistoric diffusion of metal in the Old World were quite broad. Virtually all peoples of Eurasia used metal before the advet of history. Only the most primitive of Africans—the Pygmies and the Bushmen—did not do their own ironworking. Southeast Asian metallurgy, itself of probable Indian origin, spread over Indonesia. Beyond there, however, there are only sporadic hints of metal-using peoples in western New Guinea and then nothing more to be found in the Oceanic area or in Australia.

In the New World metal was confined to the area between central Mexico and Bolivia or northern Chile. To the north of this region there were only the uses of native metals discussed earlier, strictly nonmetallurgical, it would seem. Further it should be noted that the metallurgists of the New World made no use of iron but stayed with the silver-gold-copper-tin group. These softer metals were used more for ornaments than for tools so the cultural consequences of a Metal Age were different in this hemisphere.

In archaeological context, the metallurgical analysis throw light on the metal artifacts and the tools or waste products of their manufacture are examined to reconstruct working process and the source of raw materials obtained in the past. This may be done by various techniques of chemical analysis or may involve metallographic examination under both optical and electron microscope[4] techniques. Metallurgy is sometimes divided into ferrous (iron) and nonferrous (other than iron like copper, gold, silver, zinc, lead, etc.) metallurgy, showing the dominant importance of iron and its alloys and their ore too. In the case of nonferrous metals like copper, bronze, brass, gold, silver, zinc, lead, etc. such analysis may yield information about alloys, casting, cold working and annealing. But in ferrous metal, such as iron and steel, there may be information about forging, carburization, quenching and tempering are provided by metallographic analysis.

The thermodynamics[5] in controlling the use of fire in pyrotechnology[6] has been an important land mark in the process of metal technology. Pyrotechnology involving a very high temperature which was not obtained in potter's kiln, because 400^{0}C to 600^{0}C was the maximum temperature of the potter's kiln in ancient time, when clay is baked by which further water begins to be lost from the molecules of the clay cannot be replaced and the clay is turned to pottery. Somewhere above 1000^{0}C the particles begin to fuse, but such temperatures were beyond the control of most early potters.

The purpose of firing in making pottery is to weld clay particles together. Pure Kaolinite [$Al_2Si_2O_3(OH)_4$] melts at 1770°C, far beyond the reach of early kilns. But with the inclusion of impurities in the clay which act as a fluxes[7] , the clay particles will sinter together at a much lower temperature i.e. 400°C to 600°C for most of the ancient pot clays. But kiln temperatures improved from neolithic to chalcolithic which is ample testimony to the fact that extraction of metal from its ore and temperature increase in the potter's kiln could be interconnected. The smelting point of copper from its ore is 1083.4°C; iron from its ore is 1535°C; gold is 1064.4°C; silver is 962°C, tin is 223°C and zinc is 419.58°C. Thus theoretically the melting point of only tin and zinc was in the catch range of ancient potter's kiln, while copper, silver, gold and iron were beyond their reach. Here tin can easily be obtained from its ores, but since it is a very soft metal, so it was not much useful for making implements. The pure zinc is also very soft metal like tin, but its extraction is most complicated when we compare it from both ferrous and non-ferrous metals, because it produced by volatilising[8] of zinc (boiling point 907°C) first into vapour state and then this zinc vapour condensed into solid metalic zinc. By this technical complication, in the history of metal technology zinc was invented very late as an alloy of brass, which is basically a copper-zinc alloy containing more than 50% copper. Like zinc, independent tin was not much useful in the past but its alloy like bronze (an alloy of copper and tin) the optimum proportion being about 90% copper and 10% tin. It had many advantages over pure copper; it had a lower melting point, was harder and, above all, was easier to cast without flaws. Alloy is a substance consisting of two or more elements of which at least one is metal. It may be compounds solid solutions[9] or mixtures of the components. Most pure metals can be strengthened by small additions of other metals or non-metals. The most striking example is steel which is initially an alloy of metallic iron and non-metallic carbon. Pure iron has a tensile strength (the stress in a material that is being streached) of the order of 14 tons/in^2 and a hardness of approximately 70 H_D /30. With 0.5% carbon, the tensile strength in the softened condition is in excess of 30 tons/in^2 with a hardness of 160 H_D/30.

Likewise brass having 50% zinc giving tensile strengths upto 50 tons/in^2, while 40% zinc and 60% copper known as muntz metal[10] representing a duplex brass with a tensile strength upto 25-30 tons/in^2. Thus both bronze and brass are considerably harder than pure copper, likewise steel is harder than pure iron. The hardness of pure gold is 20 H_D/30 and tensile strength is 7 tons/in^2 but when it alloyed with copper or silver then its hardness increases to 120 H_D/30. The purity of the resulting alloys is described as in carats[11] (commonly 9, 18, and 22) where these are the number of parts of gold (by weight) out of 24 parts. The 24 carats unalloyed gold is very soft, so its alloying is a must for making ornaments. Apart from hardness in alloy, their melting point is always less than the melting points of their ingredients, thus the discovery of alloy was initially not intentionally but it was due to not obtaining such a high temperature in the furnace by which pure metal can be melted and extracted. Because such a high melting point temperature where beyond the reach in early pyrotechnology e.g. Pure iron melts at 1535^{0}C a temperature inaccessible before the 19th century A.D. Hence in antiquity iron could be reduced from its ore in molten form only if during smelting process it absorbed carbon from a charcoal fire, thereby forming an alloy with a lower melting point. When the carbon content reaches 3 to 4.5% the melting point falls to about 1150^{0}C not much higher than the melting point of copper (1083^{0}C) or the temperature of a good pottery kiln. Pure copper melts at 1083^{0}C, copper with 8% tin melts at 1020^{0}C. With 13% tin it melts at 980^{0}C and with 25% tin its melting point decreases to 800^{0}C shows the pyrometallurgical advantages of alloying. Thus the beneficial effect of alloying has been taken advantage of is lowering of melting point and increase in fluidity of the metal which increase the hardness and strength of the metal.

In nature mineralization is the process of oxidation[12] by which valuable minerals are introduced into rock, resulting in an ore deposit either actual or potential. Sometime, it is formed by the process of fossilization in which inorganic materials replace organic feature by slow oxidation in the crust of the earth since sometime in the geological past. But ore is a natural mineral from which useful economic substances can be extracted by the process of smelting. Thus all ores are minerals but all minerals are not ores.

The chemistry of minerals/ores can be defined as a crystalline matter and the atoms of chemical elements entering into their composition are found to be in regular distribution in points of lattices and maintained in the state of equilibrium by the forces of mutual attraction and repulsion. The nature and value of these forces are determined by the properties of interacting atoms of chemical elements. The formation of any chemical compound including minerals and ores may be imagined as a convergence of free atoms till at a certain distance between them is reached at which point the force of attraction and repulsion are brought into a state of equilibrium by the process of oxidation. The chemical nature of minerals and ores are determined by the main or species forming elements the content of which usually amounts to units and tens of per cent by mass. In minerals they are found to be in a particularly quantitative relationship multiple to the prime numbers. Numerous minerals have a constant chemical composition, for example, quartz SiO_2, chalcopyrite $4(CuFeS_2)$, chalcosiderite $Cu(Al,Fe)_6(PO_4)_4(OH)_8(4H_2O)$, chrysocolla $CuSiO_3.2H_2O$, bauxite $Al_2O_3.3H_2O$, hematite Fe_2O_3, magnetite Fe_3O_4, siderite $FeCO_3$, kaolinite $Al_2Si_2O_5(OH)_4$, vivianite $2[Fe_2(PO_4)_2.8H_2O]$, xenotite $CaSiO_3.2H_2O$, etc. However, minerals of variable composition, with various content of separate or all species forming minerals as well as foreign admixture are much more widely spread. For instance, in native gold[13] the presence of several percent of silver and copper the content of which in samples from different deposits is difficult. This is due to metals available in the ground of the earth surface in bulk metallic form by meteoric processe.g. meteoric iron. It usually contains a moderate amounts of carbon, chromium, cobalt, copper, manganese and phosphorus. Among the species-forming elements hydrogen holds the leading place. It enters into the composition of over 2000 mineral species. In minerals hydrogen almost invariably forms oxyhydrogen groupings of various types. The heating of minerals containing hydrogen results in extraction of water. The separation of water takes place within one or several intervals of temperature which shows various degree of strength, and consequently, various types of bonds sustaining water molecules. So the formation of all minerals (native, sulphides, halides, oxides, hydroxides, silicates, borates, carbonates, nitrates, phosphates, sulphates, molybdates) in nature is due to long process of oxidation thus smelting of metals from ore is just a reverse process of it means removal of oxidation effect i.e. reduction.

Because metallurgy includes the ability to smelt metals from ores, one must start with the ores at the mines. We have no sure way of knowing how the first smelters became aware of the relation between ores and metals, since they bear little resemblance to one another. Further, a given metal may come from a number of ores, each quite different in appearance. Whatever it was that directed the attention of early metallurgists, they obtained their ores from the surface. Open pit mining was the most common form employed, but veins of ore were followed for short distances into the earth.

Mining was not far different from the quarrying which had been a part of stoneworking. Ores were dug out with many of the same tools: antler picks and rakes, stone mauls, scapula hoes. Few metal tools were used until iron became abundant; tools made of softer metals were ineffective and the later, better metals were needed initially for such important purposes as weapons. Techniques such as fire-setting and wedging were used to crack and loosen the ore bodies.

Open pit mines, found worldwide, were extensively used in prehistoric India and Africa in those same areas noted today for mining. That many of the Indian and African pits have been backfilled has puzzled some observers. It is suggested here that the entire pit was never open at one time, but that the mining proceeded as a trench moving laterally. This explanation is supported by the Africans' motor habits using the hoe, their common digging tool. The user stoops over, chops away in front of himself, and throws the earth back past his legs. In this fashion the ore would be collected and removed in baskets, and the waste thrown behind to "backfill" the pit.

The habit of following ore veins downward from the surface led to the digging of bell-shaped pits—shafts which were widened below ground level to maximize ore recovery. Such pits, due to their undercuts, were hazardous to miners and to the general public; people or animals who tumbled in could not get out without help.

Because there was no sure way of locating totally buried mineral deposits, though dowsing was employed in medieval times, shaft-and-gallery mining came later and was not generally found among primitives. Only the richest ore deposits were exploited, in part because there was no pressure to use low grade ores, and in part because the smelting processes would not handle any but the highest grades.

The expansion of mining and smelting led to serious deforestation in some areas as trees were cut to provide timbering for the mines to prevent collapse of extensively excavated chambers. Had the miners been content to remove less ore there would have been less need for pit props. Wood was also needed to make the charcoal with which smelting was conducted. Until the development of new techniques it was not possible to substitute coal for this purpose.

Once excavated the ores were crushed and washed. Panning was employed, for example in India, Africa and South America, to concentrate the metallic ores. The greater weight of the metalliferous portion caused it to remain in the pan while the lighter impurities floated away. Though mining was usually done by men, women and children joined in the transporting and cleaning of ores.

Smelting was generally done by reduction of metallic oxides in a furnace. Hot gases containing carbon combined with the oxides to carry off the oxygen in carbon monoxide and dioxide compounds. If the ore to be smelted was not an oxide, but a carbonate or sulfide of the metal, then it might be subjected to preliminary roasting to convert it to that form. Roasting was done in the open air by piling the ores on a large fire by which it converted into oxide ore.

The charge of a smelting furnace consisted of layers of fuel (usually charcoal) and ore, with the possible inclusion of a limestone or other flux which would combine readily with impurities and hold them separate from the fluid metal. Once these materials had been placed in the furnace, it was set afire, the draft adjusted to produce a proper blaze, and the whole thing tended for hours or even days until combustion-reduction was complete. Though fuel, flux, or even ore might be added during this time, none of the primitive smelters ran on a continuous production basis. Ultimately the fire died down, the furnace opened to let out the slag, and finally the metal removed from the very bottom. The furnace was then cleaned, possibly repaired, and recharged for another run. Production from such operations must have been low by modern standards. Recent efforts to bolster the Chinese economy through the encouragement of similar crude smelters seems to have netted no more than twenty-five tons of iron per smelter annually. The size of pre- and protohistoric metal hoards suggests that early figures were even smaller.

Primitive furnaces ranged from single pits to elaborately equipped structures. The pits were rare because they worked rather poorly; most were about two to three feet in diameter and in depth. Some had draft-inducing auxiliaries, discussed below. The most common furnace was one built on top of the ground and about the same size as the pit furnace. Some of the large African furnaces were made by eviscerating ant hills. The tightly cemented walls withstood heat well and could be used repeatedly. The taller furnaces, ten or more feet high, sometimes had sheds or roofs built around them to shelter the work crew during the long hours when it was necessary to stay with the lighted furnace. The ground level furnace had a trench dug into its base into which the slag could run.

The draft for the furnace was more commonly forced than natural, though some arrangements relied on prevailing winds.The pit or trench furnace had a trench directed from its base into the wind in an effort to fan the fire. A hillside furnace is known which had a tunnel funneling the wind into the lower part of the furnace.Neither of these arrangements is as reliable as an artificial draft furnished by bellows.

The many types of bellows used in furnaces and forges doubtless supplanted the use of fans and the blowpipe in some cultures. New World cultures do not seem to have had indigenous bellows. The bag bellows is one of the simplest, and possibly oldest patterns. A bag has one end arranged on that it may be opened and closed by the hand of the operator. The other end is fastened to a tube leading to the fire. As the operator raises the bag, he opens the top so that it may fill with air, then squeezes the top closed and presses down to force the air out the other end into the tube. This style of bellows, the most widespread form of the device, was known in the Old World from southern Asia across the Near East into Europe and into much of SubSaharan Africa. Derived from it is the concertina bellow which uses a rigid base and a series of internal rings to hold a longer bag in shape. The valving is still handled through the top. Bellows of the type recently common in western Europe are evidently derived from the concertina type and so ultimately from the bag bellows.

The Sub Saharan Africans made considerable use of several varieties of pot or drum bellows which consist of a base chamber of wood or pottery to the top of which is fastened a loose diaphragm of leather. In the center of the diaphragm is a loop or stick handle by which the operator raises and lowers the skin. This bellows obviously lacks a valving system, but it operates effectively by reason of its use with a Venturi tube. The air expelled by the bellows draws free air through the Venturi tube toward the fire. When the diaphragm rises and draws air through the outlet tube, most of the intake air comes from free air rather than back through the Venturi tube.

In the Southeast Asian culture sphere the common type was the piston bellows. A loosely fitting piston runs in a vertical tube, with a cloth or feather edging forming a seal to the inside of the cylinder. As the piston is raised the seal collapses inward permitting air to pass to the underside of the piston. On the downstroke the seal is pressed outward against the cylinder wall and the air is forced out the bottom of the cylinder into a tube leading to the fire. The valving system is virtually the same as that on our present day tire pump.

Bag bellows, the derivative concertina bellows, pot or drum bellows, and piston bellows were usually operated in pairs. The smelter's or smith's assistant operated one of the pair with each hand. For a forge a pair usually sufficed, but a smelter's furnace might need several pairs. The tube through which the air is carried into the fire is known as a tuyere. Tuyeres, with or without forced draft, are commonly used in multiples.

In the Far East—China and Japan—is found the box bellows which is far more sophisticated than any type mentioned above, even more than the European bellows. The box bellows consists of a long rectangular-sectioned chamber (the cylinder) in which there is set a movable partition (the piston) dividing the chamber end from end. Each end wall has an intake valve and in the side wall near each end is a port communicating to a passage located alongside the main chamber. At the midpoint of this passage is a lateral outlet with a swinging valve which alternately blocks one end of the passage to prevent the existing air from recycling to the other end of the cylinder. The piston is mounted on the end of two parallel shafts which protrude through the end wall of the cylinder. A crosshandle gives the operator his grip. As the piston is moved in the cylinder, air is forced alternately from the ends of the cylinder. This double-acting air pump is used singly to supply draft to a furnace or forge.

The techniques of metalworking differed according to the metal being used and the results desired. Cold hammering was widely used with the more ductile metals such as gold and silver. It could be pursued with other nonferrous metals, but continued working made copper and bronze brittle. If the desired shape was not achieved before embrittlement, then the metal was annealed by heating and rapid cooling. It is possible that annealing was the first use made of heat in the whole metallurgical complex, preceding smelting or welding. Hot forging of metals became much more common when iron came into use, since only a limited amount of cold working of steels can be done. Heating, followed by slow cooling, annealed steels; rapid quenching of the heat made steels harder, tempered them.

Most iron used by man is actually alloyed into steel of some sort. When iron ores are smelted with carbon fuels, the carbon in the furnace combines with the product to make a steel. Mild steels, which we commonly refer to as iron, have 0.05 per cent to 0.30 per cent of carbon. Steel for today's tools runs about 0.75 per cent to 1.0 per cent carbon, while very hard items like files run to 1.25 per cent carbon. (Other metals are additionally alloyed in modern steels). Since the differences in carbon content between some radically different kinds of steel are quite small, the metallurgist must exercise careful control. Evidently primitive smelting was at times a rather hit-and-miss affair.

Since most primitive furnaces were not capable of very high temperatures, they did not produce quantities of molten iron. Instead the iron formed as droplets, so-called point melting, in the furnace and trickled down to form a spongy mass, the bloom, in the base of the furnace. After the smelting process had run its course, the bloom was dragged from the furnace and possibly broken into smaller pieces. This metal was heated and hammered repeatedly to control the quantity and distribution of carbon which, depending upon the treatment, could be added or removed at this step. But for most nonliterates, the low furnace temperatures precluded the casting of iron into molds.

The drawing of wire is basically a forging processeven though it is done cold among simpler peoples. A rod of metal was hammered down at one end until it would pass through a hole in a drawplate or die. The reduced end was held in a clamp and the rod pulled through the hole by force. The process was repeated, using successively smaller holes in the downplate, until the intended wire size was reached. Drawing was done best with the more ductile metals, but others could be annealed between passes through the die. Various kinds of levers and simple windlasses were devised to supply the considerable force necessary. Wire drawing was known in SubSaharan Africa and in a belt from Europe across southern Asia to the Far East.

The casting of metals was a widespread technique with several methods known in both Old and New Worlds. The nonferrous metals lent themselves well to casting, but, as noted above, iron casting was late and restricted. The simplest kind of casting is done into an open mold to form ingots (some used for money), rods and more elaborate shapes. Some open molds were simply earth cavities, but others were formed in clay or cut in stone. The stone molds, usually in soft limestone or steatite, could produce castings of some detail and precision. The open side of the mold left the casting unfinished on that surface, so some later finishing was necessary. Many ornaments, such as brooches, buckles, even bracelets, may be made in this fashion.

An elaboration of the open mold involved the addition of a flat cover to close the open side. While this does no more than confine the molten metal and control the ultimate thickness, the addition necessitated passages for the admission of the metal and vents to purge the evolved gases. The closed mold raises problems of premature chilling of the metal add resultant incomplete filling, but preliminary heating of the mold reduces the heat loss and aids the flow.

Split molds are closed molds with a part of the casting cavity in each mold section. Often the molds are in two pieces, mirror images of each other. Molds of three or more pieces may be necessary for the casting of complex forms, such as socketed axes. Some means must be furnished—like aligning lugs—to make the mold pieces fit together properly.

Olds usually are treated with a parting agent, usually powdered charcoal, which keeps the metal from sticking to the mold. Properly designed and used, an open or lidded mold or a split mold may be used many times. Many identical items can be made in this fashion before the mold needs to be replaced.

Casting in closed molds always leaves its mark on the product. Some metal leaks into the crack between adjoining parts and solidifies there. The casting freshly removed from the mold, has a thin projection of metal, called flash or fin, which is trimmed off in the process of finishing and polishing.

If this line of protruding metal is not completely removed, it may be possible to detect the lines of juncture in the mold.

Casting by the lost wax (cire perdue) method was evidently an independent invention in each hemisphere. There are ethnographic records of the method in Europe, West Africa (by diffusion from Europe), and Indonesia. By implication of archeological artifacts the method was known in the Middle American area and in Peru.

In essence, the lost wax method involves the forming of a wax model of the object to be cast. Over the model layer upon layer of fine clay are placed until a mold is formed. In the forming process passages spures must be left for the metal to be poured in and vents left for the gases. The model and its covering (the mold) are heated to melt out the wax. Then the molten metal is poured into the mold. Once it has cooled the mold is broken open and the casting removed. Major advantages of lost wax casting include the ease of formation of model and mold and the ability to make castings of complicated shapes; one disadvantage is the need to make a new model and mold for each item.

There are many variations possible within the general framework of lost wax casting. The model itself can be cast in order to make repeated casting of the same thing. Formed of several pieces of cast wax, the model can be assembled with yet other pieces of wax into a coherent whole. Any modeling materials which can be burned out of the mold can be used; some West Africans used insects and plants as their models.

When design considerations call for a bulky casting, such as the body of man or animal, then a hollow casting is desirable because its thinner wall will facilitate the escape of heat from the freshly-poured casting and conserve metal as well. A core is made which is the size and shape of the ultimate hollow. Over this is modeled a thin coating of wax to equal the thickness of cast metal. Pins of an unobtrusive metal are set into the core. They pass through the wax and protrude into the walls of the clay mold which is now formed over the wax. The whole mold is heated to drive off the wax, and the metal is poured. The mold is removed, the pins trimmed flush with the metal surface, the core broken out of the interior, and the casting polished. The entire process must be repeated with each article cast.

Though casting in split molds and other more refined methods are common today, lost wax casting remains the choice for works of art and for experimental, one-of-a kind items e.g. dansing girl of Mohenjodaro.

The anvils and hammers were often augmented with forms in stone and metal which served as swages. A special anvil might be grooved or have cavities into which the hot metal was hammered. A separate swage might be held atop the metal which in turn rested on the anvil. The top swage was then struck by the smith. Swages and similar forging forms were, of course, closely related to the products to be made. The ribs in many African spear heads were formed with the use of swages.

Chisels and punches were uniformly of metal because stone chisels, while adequate for woodworking, were not suitable with metal. Tongs were of the bent-loop type rather than hinged and if made of green wood rather than metal were adequate and cheap. In the absence of tongs green sticks were used; in fact, it was common practice in Africa and India to insert a part of the metal being forged into a green stick which then served as a handle until forging was completed. The same kinds of things which were used to handle hot metal were also employed to manipulate hot crucibles of molten metal or heated molds for casting.

Aside from the drawplate and any windlass employed, the only special tools for wire drawing were clamps to hold the wire. These clamps, of the loop type like common tongs, had a sliding ring on which the pulling force was exerted. Thus, the harder one pulled, the tighter the grip on the wire.

One of the first alien tools used by many primitive metal workers was the file, highly prized by these artisans. Not only did their use greatly facilitate the shaping and finishing of metal goods, but their metal was substantially harder and better than most nonliterates were able to make. Bits of wornout files can be found incorporated into many artifacts of the early contact period in different parts of the world.

Much of our knowledge about the metallurgy of nonliterate peoples has been systematically summarized from direct field observations of ethnographers, travellers, and others who have described techniques which have persisted into recent times. However, another substantial body of data is of archeological derivation—from the metallurgical analyses of the products. These have been made for years and tell us something about techniques of manufacture, ore sources, sophistication of work as reflected in degree of consistency of product, and the diffusion of processes. Some archeological contexts have yielded the goods of the metallurgists themselves—their furnaces, crucibles, tools, and molds. The molds have occasionally been used experimentally thousands of years after origination, to cast further products of the original type, and thus to illuminate our understanding of these ancient technologies.

Smelting is the process of separating metal from ore by the process of reduction which is usually effected by the presence of carbon or a proper ratio of carbon dioxide and carbon monoxide and a flux is added to form a fusible and liquid slag[14] which will float on the metal and collect the impurities which it is desired to remove. Some metals may melt whilst being smelted, but this is not necessarily part of the process. The main chemical reaction in smelting most of the ores used in antiquity is that of reducing a metal oxide. If the ore was not already in the oxide form, then it was converted by a preliminary process. Most ancient copper probably came from oxide ores (cuprite—Cu_2O, melaconite—CuO) was reduced could have been as low as 800°C—the kind of temperature reached in a simple hearth or kiln. In many cases however, the copper was melted into ingots[15] during the smelting operation. This requires a temperature of some 1083°C, which would have necessitated the use of furnace or hearth assisted by bellows. Furnaces is a refractory lined chamber in which temperatures above 300-400°C can be produced, because equipment for lower temperatures is called an oven. Blasting by bellows increases the temperature of the furnaces from 400°C to maximum 1000°C, and whose maximum limit is subject to the nature of fuel and air for blasting. In Indian context '*Bhastra*' word for air bellow was used in *Satapatha Brahmana* (XI.2.1.16) which conform the earliest textual evidence of use of air bellow in blasting furnaces during *circa* 1000-500 B.C.E. Technically careful control of the amount of air entering the furnace would be required for successful smelting. Similar, or lower temperatures would be required for the smelting and melting of gold, silver, lead and tin, but melting point of iron is 1540°C beyond the range of simple furnace. Iron could, however, be smelted. The reduction of iron oxide ores can take place at 800°C. If the temperature is increased to some 1150°C, the slag melts can be drained off. This leaves a spongy bloom of iron and remaining slag inclusions. The slag within the bloom must be removed by forging.[16]

It is clear that no people could enter in metal age (Bronze → Iron →Brass) from lithic age unless their ores were in fact available in their environment. Hence the Eskimos of Arctic region[17], who have no metals available to make into weapons or ornaments, still use stone, ivory (from various tusks) and bone. In short their environment due to non –availability of ore does not permit to enter into metallic age. Thus the primitive races or people have evolved ways of life which are often well adapted to the environments in which they lived. Geologically India has an impressive deposits of both ferrous and non-ferrous ores. The remains of ancient mines and smelting furnaces located in the vicinity of modern mining centers as well as the evidence of common occurrences of slag and few metal-working places presented by the unearthed ancient sites reflect man's earliest endeavour to recover and utilise these metals in the process of cultural development. The modern civilization and cultural progress of mankind has been influenced to a large extent by mining of metallic ores and their extraction; probably some 9000 B.P. According to famous French mining historian Simonin[18] "The intrepid coal miner, whose advent is of but recent date; the veteran of the mineral world—the metal miner—whose origin dates from the dawn of the history; and the patient seeker after gem.... Everybody has thought, at some time or other, of the important part played by metals in the life of nation. From the most precious to the sasent, from the rarest to the commonest, they all play a special part; and material progress is only affected by their means. The history of metal constitutes the true history of inventions and labour." Mining began some 45000 B.P. with the Palaeolithic man as evidenced by flint implements found along the bones of early man. At first, the Prehistoric man recovered stone raw materials from the surface explorations but by the beginning of the Neolithic period (*circa* 10,000 B.P.) he had progressed to underground mining. However, the oldest datable underground mine is from Bọmvu Ridge in Swaziland which is 40,000 B.P. and the mining material was hematite.

Panini in his *Astādhyāyī* recognize that the word *khan* (means mine) had appeared in the *chandas*. The commercial importance of mining in his time is revealed from his discussion on *akarika* (4.3.75). The word *bhrastra* (6.4.47 and 4.2.36) meaning one who roasts suggests that the sulphide ores were probably roasted before smelting and air bellow was used in the furnaces probably to smelt iron from its ore. Mines were considered to be of supreme importance towards the economic welfare of the society and defence of the nation against external aggression as per *Kautilya Arthasastra* (2.12.37 and 2.14.25). The Superintendent of Mines was supposed to be an expert in mineralogy, smelting and colouring gems (2.12.1) and to establish factories for caustic alkali (2.12.27). The Director of Mines or *Akaradhyaksa* was supposed to be an expert in mineralogical veins in the rock (2.12.1; 2.24.10). He could open new mine or tap, the residual ore from old mine, detected by the marks of drons, crucibles, coal and ashes (2.12.1). He describes the southern country (*Daksinapatha*) as yielding conch-shells, *vajra* or diamonds, gem stones, pearls and gold. He further states that the southern route has many mines (7.12). Allchin summarizing this point analyses the location of Asokan edicts *vis-à-vis* the mines as follows:

"The minor Rock Edict at Maski is actually on the gold-field; the two Koppal edicts are no more than 30 miles from the Gadag band gold-mines; the three at Brahmagiri are equally near to several mines in north in Mysore and Sandur; while the other two sites of inscriptions at Yeragudi and near Pattikonda are in the center of country long famous for diamonds."

He further suggests that the town of Suvarnagiri mentioned in the Brahmagiri edict could be identified with the golden hill of Maski. Thus the emergence of different dynasties was not merely a coincidence but a historical necessity born out of several factors of which principal one must have been the regional exploitation of the natural resources of the land. From the periods of foreign travellers like Megasthenese (*circa* 300 B.C.E.), Plinny (77 A.D.), Fa Hien (5th century A.D.), Huen Tsang (7th century A.D.), Al-beruni (11th century A.D.), Travernier (1665-69 A.D.) information on ancient mining and use of metals and alloys in India in ancient times is found.

Technology may be seen to consist of a series of coordinated techniques or particular methods of reaching a technical goal—the molding of pottery or the tempering of metal. Though technologies in the narrower senses are usually distinct, their techniques may be shared by several technologies. Burnishing, for example, may smooth the surface of a pot as well as that of a metal piece. Casting is likewise a shared technique found in both ceramics and metallurgy. However, similarity of name should not be the basis for judgment of technical sharing, as the process of tempering clay for ceramics cannot be equated with tempering in metallurgy.

A technological pursuit requires an input so that a situation may be created, procedures followed, and a result obtained. The three major elements of the technological input are knowledge, resources and labour.

A group's knowledge has a great bearing on its technological capability and behaviour. Though innovation occurs, action stemming from innovation constitutes a small part of the behaviour at any given time. Most technical knowledge is a legacy from earlier times. It is only in periods of great innovative activity that any substantial portion of the effective knowledge of a generation is new. Estimates have been made of the continuing elements, the legacy, in some limited technological spheres. Despite their rather impressionistic character, the estimates tend to confirm the essential continuity involved. They also indirectly support the notion that people rarely rise above the general level of technical knowledge of their time. What is known at a given moment is, then, a strong limiting factor in technological behaviour. However, pure knowledge, that of science, frequently runs well in advance of practical knowledge, that of technology. The lag, which always exists to some degree, is variable. It is influenced by the rate of scientific advance, the ease of communication between science and technology, and the state of the society.

Resources constitute the second element of the technological input. Their availability and the knowledge to exploit them are factors in the end result of technological activity. If the resources are not present locally, a society imports them (at what cost?) or does without. However, the presence of resources is no guarantee of their utilization. The group may lack the knowledge prerequisite to exploitation or it may, for some reason, decide not to use them. The absence or scarceness of certain resources may either inhibit some lines of technological development or may, less commonly, call forth remarkable ingenuity in the use of substitutes or the economical use of a scant supply. Whatever the circumstances, there is rarely a direct correlation between the presence of resources and their utilization.

Labour, like the other elements of input, is subject to many ancillary considerations. Both quantity and quality of labour have their importance. As to the former, some undertakings are difficult or impossible if the manpower is not available or cannot be organized toward the task. The disclosure, through archeological investigation, of large public works is usually taken as prima facie evidence of some social complexity. However, a simple census could not reveal the labour force applicable in a given situation. Many of those present may not work toward the stated goal for social or even ritual reasons. Conversely, the society may be able to call on the labour of those not normally in its work force when an exceptional task is at hand. One must remember, additionally, that even in the simplest societies there is often a sharp division between the work of men and women which precludes any of the one sex engaging in the work of the other.

Thus the invention of metal technology affect the socio-economic changes in the past and present too. The less technically advanced society is primitive when we compare it with more technically advanced societies and, of course, under the impact of advanced technology, even the primitive communities are beginning to enter a new state of development. Whilst new disoveries are by their very nature quite original e.g. the discovery of how to make bronze (the first alloy known to man) from its constituent metals of copper and tin, was something entirely novel in man's history, new inventions are for the most part a combination of old elements—that is, they are dependent upon already known principles. At follows that the more of such elements or principles existing in a particular culture the greater will be the number of possible new inventions and its diffusion. The diffusion is the spread of a technique or cultural trait or a complete way of life from one area to another. This can take place through the movement of people or through the spread of ideas (sometimes known as stimulus diffusion).[19] It is clear that diffusion has often taken place in the past and that it has sometimes been a potent force for change. However, general interpretative frameworks based on diffusion are now less popular than they once were in the case of metallurgy. The theory of diffusion have been now challenged, partly as a result of chronometric dating and availability of ores and partly on theoretical grounds, which support the possibility of independent invention in more than one area.

Thus the technological change does not differ markedly in character from cultural change, because technology is a part of culture. There are a few specific circumstances which influence the particular without affecting the general. These stem from the technical nature of technologists rather than from any exceptional factor which exempts them from consideration as a part of a culture. Cultural change has been a focus of attention, but comparable interest has not been displayed in cultural persistence which is the other side of the same coin. The presumption is that persistence is the absence of change. The equation could be read as well as in other direction. Therefore, under the rubric technological change we should consider equally the topic of technological persistence, for together they form an illuminating pair.

In the various early Indian text albeit very meagre mention of metallurgical extraction theories has been made. Even then metallurgical output like ferrous and non-ferrous metals like gold, silver, copper, tin, bronze, zinc, brass, etc. were found much earlier than the invention of writing in colossal quantity in India. How Indians were extracted these metals from their respective ores without knowing its scientific theory of extraction from the ore i.e. its reduction is still a big question mark? The other question mark which may be underlined is sheer the increase in temperature of the furnace does not complete the process of smelting, because the chemistry of smelting of the metal from its ore is highly complicated e.g. the melting point of iron is 1540^{0}C but at such a high temperature if an optimum ratio of $CO:CO_2$ is not mentioned in the furnace, iron cannot be extracted from its ore, howsoever the temperature may increase? If the optimum ratio is not maintained from 8 to 9 hour then reduced metallic iron from the ore is reoxidised into its ore form. How the ratio of charcoal and ore was developed in the furnace? In modern time the smelting of iron can only be possible when calorific value of the furnace is 800-1000 K.Cal per cubic meter. How this high calorific value for 12 hours was maintained in the furnace in the past? This may be substantiated from the fact that extraction of iron was invented in India in *circa* 2000 B.C.E., while its theory was developed in the 19th century A.D. How the corrosion resistance in iron in the form of cementite (the compound of Fe_3C) were developed in ancient Indian iron? Was it a result of empirical practice, without knowing its theory? Likewise zinc was extracted by the process of vapourization during post-Gupta period in the region of Gujarat and Rajasthan, while its theory and practical extraction in Europe was commenced during 18th century A.D. The same system (without knowing theory the practical application by empirical process) is applicable in Bell metal of China which is usually a tin-bronze alloy containing 20-25% tin sometime a little zinc as deoxidiser which balance copper whose antiquity at Erliton in China dates back to mid 2nd millennium B.C.E.

Was it due to the theory of 'learning by trial and error'? Because learning is that behaviour acquired during the living organisms life time as a result of experience. The more complicated types of learning are found only in those animals with elaborate nervous system like modern man (*Homo sapiens sapiens*). The trial-and-error learning occurs only when spontaneous behaviour of a living organism accidently produces a reward. But insight learning also pay an important role in the development of technology. The cerebrum of the brain of modern man which is the center for memory, hearing, sight, speech, voluntary muscle action, thinking and reasoning is responsible for development of learning by which innovation in technology taking the proper shape. The technological ability of man are a part of his culture and are learned and transmitted outside the genetic realm. The nature of technological innovation are not static but are dealing with a dynamic situation. However, any activity which uses or controls material objects is linked to some technology through those objects.

Social structure and technologies are similarly related. The social standing of workers and technicians, especially as these come to be a special segment of the society through the development of so-called true division of labor, may either encourage or inhibit technological pursuits. In earlier times, and to a lesser degree today, technical knowledge and technical facilities were transmitted in family lines or within some larger kinship unit such as the caste. The recruitment of technicians even in the absence of direct inheritance, often lay within a group of kinsmen. Social status in many societies is correlated with the ability to command, accumulate and distribute material goods.

Though the bond is not as evident, religion and technology are also linked. At various times and places, as in dynastic Egypt and medieval Europe, a large proportion of the technological output has been directed toward religious concerns. Pyramids and cathedrals are the tangible fruits of these enterprises. Less obvious are the techniques and other advances which resulted from the pressures of the times. Most of these were applicable to other pursuits. In addition, religious ritual has been practiced to insure the success of various technologies. Without proper care for the supernatural world, the efforts of man were likely to fail or fall short of complete success.

Because technology is a part of culture, there may be applied to it all of the concepts which deal with culture generally. These may be used, perhaps more than they have been in the past, in the analysis of technology. All the elements of theories of cultural dynamics are applicable, among them cultural evolution in its various forms, diffusion[20] and its adjuncts, and functionalism of various persuasions. However, this statement is more an article of faith than an exposition of the approaches actually used. Some have yet to be applied to technological analysis.

Notes and References

1. Alkaline is an acqueous solution with a pH greater than 7 and turns litmus paper blue.
2. Cation is the ion that moves towards the cathode in electrolysis; as cathode is negatively charged in electrolysis, cations are positive ions.
3. Ion is a species that carries electric charge. There are two types of ion. Positive charge is carried by positive ions (also often called cations) which have fewer electrons than their parent atoms. Negative chage is carried by negative ions (also often called anions), which have more electrons than their parent atoms.
4. Electron microscope is an instrument that creates a high-magnification image of an object by employing a beam of electron (rather than light as in conventional optical microscope). These instruments can attain a resolution of the order of 10^{-3} A° compared with a limiting resolution for the optical microscope of the order of 0.2 A°.The material which can be examined under electron microscope must be able to withstand high vacuum, it must be sufficiently thin to be transparent to electrons and it must withstand the heating effect produced by electron beam. Specimens for transmission electron microscopy therefore consist either of very thin films of the material to be examined, films used to support small particles, or thin film replicas of the surface to be studied.
5. Thermodynamics is the scientific study of the transformation of energy by heat and work. It also governs the conversion of energy from one form to another, the direction in which heat will flow, and the availability of energy to do work.

6. Pyrotechnology is the process metallurgy involving high temperature i.e. above 300°C, because in normal thermometers measure the temperature upto 300°C, but above that pyrometer is used in measuring the temperature.
7. Flux is used to describe a material employed to lower the melting point of more refractory materials and to protect underlying metal from oxidation or when zinc chloride is used to clean and protect surfaces during soldering etc; or when borax is used for similar purpose in brazing.
8. Describing a substance that can easily become a vapour.
9. Solid solution is a metallic alloy in which the microstructure consists of similar crystals with no evidence of the separate existence of either alloying element, or of intermetallic compounds. Solid solutions may be substitutional if the atomic radii of the constituents do not differ by more than about 12%. If there is a big discrepancy in size factor they may form interstitial solutes.
10. Muntz metal is a 60:40 type brass of which the composition varies from 58 to 61% copper, often with upto 1% lead.
11. Carat is a unit of mass, used for precious stones, equal to 0.2 g. It is also the name given to a scale used to measure the purity of gold, 24 carat gold being pure gold.
12. The meaning of oxidation has been sharpened over the last two centuries from meaning combustion with oxygen, to the modern meaning of 'loss of electron'.
13. Native means an element found in nature uncombined with any other elements. Gold and platinum are the common example of native metals.
14. Slag is the mixture of flux, impurities which floats on the surface of molten metal. It also protects the metal from contamination by the atmosphere.
15. Ingot is a solid metal casting prepared for subsequent hot working. Normally rectangular in section and tapering slightly from top to bottom.
16. Forging is a type of plastic deformation of metal by squeezing or hammering to produce a required shape. It is the process of working hot metal: iron and steel, are too hard to be cold worked and must be heated before they can be shaped. Forging was not only used to shape metal, but was also part of the refining process for iron, where the bloom produced by smelting had to have the slag impurities forged out of it.
17. Eskimo is a long standing and widely used term for Arctic hunting peoples. It is a French translation of an Algonquin word means 'raw flesh eaters'. Invit, a native

Eskimo word meaning simply 'the people', is the term currently favoured to describe the same group.

18. Simonin, L. (London, 1969), *Underground Life: Mines and Miners*, p. 522.

19. Stimulus diffusion is a special clan of intercultural transfer in which only the basic kernel of a trait is passed over. This then becomes "stimulus" for the development and elaboration on the part of its recipients. Instance of stimulus diffusion are difficult to establish because most of the criteria which are usually evaluated to make a case are lacking. An element in the occurrence of diffusion is the location of the societies in question.

20. Gordon V. Chide (1892-1957) was the chief proponent of the diffusionist view which interpreted all major developments in prehistoric Europe in terms of the spread of either people or ideas from the Nar East.

CHAPTER 2

CONCEPTS OF MINERALS AND ORES

The History of Mineralogy

The mineral is the source of elements which is forthcoming in nature as a substance with a characteristic chemical composition in usually inorganic forms and ores (mineral from which useful substance can be economically extracted) which were formed in nature from ancient times. Having emerged along side advent of man, i.e. *Homo sapiens sapiens*[1] during Upper Palaeolithic period in the later phase of Pleistocene[2], these concepts have had a long history. According to many mineralogists a mineral is a structurally homogenous solid of a definite chemical composition formed by the inorganic process of nature. This definition includes ice as a mineral but excludes coal and natural oil.

The knowledge of the mineral world began with the use of stones that were picked up from the ground and directly employed as the most primitive tools of labour and arms. As early as those remote times Man was able to differentiate stones and select those that were more appropriate in form, mass and strength. Later, with the transition to more improved forms of processing, stones acquired even greater importance since their implementation for practical purposes had undergone substantial changes. Different properties of stones were made better use of, and the knowledge of how to discover and extract them had been gained. Articles made of silicon, obsidian, chalcedony, as well as nephrite were of particularly high quality. Besides axes and hammers, people learned to use stone to produce the more refined tips for their spears, darts, arrows, knives, scrapers and other labour tools. In Indian context the lower Palaeolithic tools[3] are made on the most common and abundantly available quartzite[4] and sandstones from the sedimentary Vindhyan and Cuddapah formations and their next preference is the basaltic rocks like the Deccan trape or dolorite. There are however some instances where preference was shown to a particular raw material not locally available as instanced by a jasper handaxe from Mandsaur in Madhya Pradesh, Chert handaxe from Chirki in Maharashtra and phyllite handaxe from Hunsgi in Karnataka implying some kind of transportation. The Middle Palaeolithic tools[5] are made of siliceous material like chert, chalcedony, agate and jasper were found in the middle or upper gravel (Grovel II), overlying or resting against a coarse gravel (Gravel I) with Acheulian industry. The crypto-crystaline raw material of Middle Palaeolithic continued into Upper Palaeolithic[6] too, although quartzite and quartz were also occasionally used at Budh Pushkar in Rajasthan, Visadi in Gujarat, Bhimbetaka in Madhya Pradesh and Renigunta and Yerrgondapalem in Andhra Pradesh. In Mesolithic phase[7] except in parts of South India where very fine grained quartzite was available and used, in all other areas people switched over to chalcedony and chert for making these new tools in the form of microliths. Quartz which does not lend itself to the production of good microliths was also sometime used probably because of the limited availability of chert[8] and chalcedomy[9]—a poor quality of flint, which is a hard but brittle stone found in chalk or limestone. It is chemically a quartz, but differs in having a microcrystaline structure. This gives it the great advantage to man that it can be flaked readily in any direction, and so shaped to some desired useful form. It occurs widely, and where available was the basic material for man's tools until the advent of metal—it is the commonest stone of the Palaeolithic age in the three age system the only types of stone preferred to it where obsidian and the tougher rocks used for ground tools in the Neolithic. Flint was usually shaped by flaking, less commonly by grinding. Since ancient times stones, bright in colour and design, had been used extensively as ornaments and amulets. Mineral paints were used for rock and cave drawings, and ritual purposes. According to the data provided by the archeological finds, by the end of the prehistoric period people made practical use of approximately fifty types of rock and minerals such as: sandstone, quartzite, basalt, obsidian, silicon, chalcedony, quartz, nephrite, clay and sand, mineral pigments (iron ochre, colour clay) and rock salt.

With the emergence of ancient cities and states the malleability of metals, the mining and the processing of the ores of gold, silver, copper, tin, lead, mercury, iron, as well as precious stones become highly significant. Copper, and later, bronze (the alloy of copper and tin), and iron were used for the manufacture of arms, labour and domestic tools. Silver, gold and precious stones were employed as symbols of power, for ornamental purposes and were widely used in the trade. Since ancient times stones and clay have served as the most important building materials and when edifices were erected, it was customary to use stone that could be easily handled, such as: limestone and sandstone; while for columns and for exterior and interior decorations, for sculptors, stones of higher ornamental quality were used. They include marble, granite, porphyry, basalt, quartzite among others. Baked clay and lead were used for the manufacture of water pipes.

For centuries and millennia a substantial knowledge of how stone and metals could be used for practical purposes, as well as the understanding of the mineral world as such, as been accumulated. This is brought to light by the remains of ancient cultures, ancient mines, the fragments of ancient Egyptian papyri and other written sources. In the Chinese chronicles (20-5 centuries B.C.E.), and the Indian epoch (11-10 centuries B.C.E.), as well as in the works of the ancient writers that have reached us, we have the first generalizations concerning this kind of knowledge and copious information on stones, reflecting the state of the art that had been attained by that time, together with their concepts and systematization of minerals, as well as the properties, structure, origin and use of stones.

The ancient Greek philosopher Aristotle (384-322 B.C.E.) explained the formation of metals by their being sublimated in a gaseous state from the earth's interior, while his pupil Theophrastus (372-287 B.C.E.) gave the first description of approximately fifty minerals. In the books of the Roman natural scientist Pliny the Elder (23-79 A.D.) there are more detailed descriptions of stones, which he brought into his system in accordance with their practical uses.

In the early part of the Middle ages mining had acquired a particularly prominent stage in its development in the countries of the East, as well as in Central America, Western and Eastern Europe, where traces of ancient mining works, smelters and forging shops have remained up to the present day. It was the time when on the territory of Central Asia there was intensive mining of copper, mercury, lead, silver, tin, iron, precious stones, ceramic clay and mineral paints. The knowledge of minerals, accumulated empirically, was generalized in the works of the outstanding scientists of the Arabic East and medieval Central Asia. Minerals were regarded by them as natural substances characterized not only by physical but by chemical properties as well. And though the term 'minerals' had not yet come into general use, the concept of minerals becomes more and more concrete.

The outstanding scientist of the Middle Ages Ibn Sina (Avicenna) (980-1057 A.D.) gives the following classification of minerals: (1) stones and soils, (2) metals (melting substances), (3) sulphuric (combustible) fossils, and (4) salts (soluble substances).

In the period of the early feudalism in Europe the development of mining gains considerable ground. As early as in the sixth-eighth centuries the mining of copper, lead and silver ores takes place in the regions that are in modern Czechoslovakia and Moravia. Mining becomes widely spread throughout the whole of Europe. Gold is mined in Hungary, copper and silver in Saxony, salt in Poland, and mercury on what is now known as the territory of

Yugoslavia. In Russia, at that time, iron ores were prospected in the regions of Tula and Serpukhov, mica was mined in Karelia, and salt in Solvychegodsk.

In contrast to other part of the world, India has had a long history of mining and metallurgy. By the time of Early Harappan phases (*Circa* 5,500 B.C. to *Circa* 2,600 B.C.E.) in the region of Ravi, Hakra, Sheri Khan Tara Kai, Balakat, Amri, Kot Diji and Sothi the knowledge of copper, tin, gold, silver and lead mining was archaeologically established, and the tradition continued right through the subsequent phases of the Indian protohistoric and early historic period. Iron appeared sometime in the first half of the second millennium B.C.E. The antiquity of the use of zinc goes back to the late centuries B.C.E. or the early centuries A.D. The point to note is that these metals occurred adequately in the geological context of the subcontinent itself. There are extensive belts carrying remains of pre-industrial mining of different metals in various parts of India. Their dates are meagre, and if the nineteenth century reports are any indication, the pre-industrial mining operatons continued in several parts of the country till the middle of that century. The long history of the use of metals by archaeological evidences in India suggests that a good number of these mines have possibly a history going back to the protohistoric and early historic periods. The reconstruction of the history of Indian mining, a thoroughly neglected field among the Indian archaeologists and historians till now, can be done only through an intensive documentation and study of surviving mining remains and the determination of their dates both by radiocarbon method and the associated archaeological samples. At the same time it is surprising fact that there is not a single old mine in India which has been scientifically investigated in detail.

On the other hand the only worthwhile ancient literary reference of mining is found in the *Arthasastra* of Kautilya (*Circa* third century B.C.E.), were mines considered to be the supreme importance towards the economic welfare of the state, society and defence of the nation against external aggression.

आकारप्रभवः कोषः कोषात् दण्डः प्रजायते ।
पष्थिवो कोषदण्डाभ्याम प्राप्यते कोषभूशणा ।। (2.12.37)

The treasury has its source in the mines; from the treasury the army comes into being; with the treasury and the army, the earth is obtained with the treasury as its ornaments.

Gold, silver, diamonds, gems, pearls, corals, conch-shell, metals, salts and ores derived from the earth , rocks and liquids were recognised as material coming under the purview of mines (*Arthasastra*; 2.6.4). The superintendent of mines (खन्याध्यक्ष, लक्षणाध्यक्ष, लवणाध्यक्ष लोहाध्यक्ष) was supposed to be an expert of mineralogy, smelting and colouring gems (*Arthasastra* 2.12.2) and to establish factories for the above items and also derivatives like the caustic alkali (*Arthasastra* 2.12.27). The metallic ores had to be sent to the respective metal works for producing twelve kinds of metals and commodities (*Arthasastra* 2.12.23 & 35-36). From the mines and the metal works had to be collected by the State, thus State was supposed to have complete control over the mines and their mineral resources. The ming head or आकाराध्यक्ष was supposed to be an expert in शुल्वषास्त्र or the mineralogical veins in the rock (*Arthasastra* 2.12.1; 2.24.1). He could open a new mine or tap, the residual ore from an old mine, detected by the marks of dress, crucibles, coal and ashes (*Arthasastra* 2.12.1). In the case of some useful information about mines, gems, the informant was supposed to receive 1/6th part, and in case he happened to be a state employer, only 1/12th part of the treasure (*Arthasastra* 4.1.51). The *Arthasastra* advised the head of mines to concentrate on the more accessible mines needing less capital investment and yielding large quantities of commodities and large profit. All transactions of minerals and metals had to be executed with the State only (*Arthasastra* 2.12.19; 2.22.10). So private ming was a punishable offence at the time of *Arthasastra* and the defaulter and other criminal were forced to labour in the mines.

The *Arthasastra* is the earliest Indian text dealing with the mineralogical characteristics of metallic ores and other mineral-aggregate rocks. It recognised ores in the earth, in rocks, or in liquid form, with excessive colour, heaviness and often strong smell and taste (2.12.1).

Of great commercial importance were metallic ores from which useful metals were extracted. The *Arthasastra* did not provide the names of the constituent minerals beyond referring to them as *dhatu* of iron (*tiksnadhatu*), copper, lead, etc.

The *Arthasastra* described a gold-bearing ore which has interpreted as typical auriferous pyrite ($AuFeS_2$). We reproduce the translation of the relevant section (2.12.5) of the *Arthasastra* with interpretation in the parentheses:

> Those ores which are obtained from the planes, or slopes of mountains; which are either yellow (pyrite, FeS_2 is brass yellow with a metallic lustre), or red as copper $CuFeS_2$, or reddish yellow (pyrite often becomes pseudomorphous after limonite, $2FeO_3.3H_2O$ and then develops brown colour) which are disjoined (individual crystals of pyrite), and marked with blue lines (grain boundary of chalcopyrite, $CuFeS_2$ and pyrite).... which are marked with spots like drops of curd (quartz is the curd-white mineral), dazzlingly bright yellow as turmeric (native gold), when broken, show lines, spots or *svastikas* of fine sand (minute crystals of pyrite occur between larger crystals; when isolated, such layers readily crumble away and become like sand grains; separation of metallic gold in the form of rings has been reported—the *svastika* represents the skeleton of a gold crystal separated out of gradually decomposing auriferous pyrite or gold bearing iron sulphide); which contain globular masses (decomposed pyrite showing iron hat); when roasted, do not split but emit foam and smoke (i.e. sulphur dioxide gas).

Similarly, the silver ore described in the *Arthasastra* (2.12.6) seems to be a complex sulphide ore containing silver in the form of argentiferous galena

(AgFePbS) which is basically an ore of lead in which silver was present as an impurity: "colour of a counch-sheil, camphor, *vimalaka* (pyrite?) *sasyaka* (copper sulphate, an oxidation product of associated copper sulphide), containing lead (galena), *anjana* or antimony (as sulphide), smells like raw meat (smell of sulphur), ores which are disjoined (cube-shaped particles of cleaved galena), grey or blackish white (colours of galena, argentite or tetrahedrite); when roasted do not split, but emit much foam and smoke (galena and argentite evolve SO_2 gas, and galena (PbS) forms white incrustation of lead oxide or charcoal)".

The *Arthasastra* described the sources and the qualities of good grade gold and silver ores (2.13.3-5 & 10-12). The quality had to be tested through streak on suitable stone-plate (2.13.17-27). The practice is still prevalent in Indian goldsmiths.

Copper ores were stated (2.12.12) to be 'heavy, greasy, tawny (chalcopyrite: $CuFeS_2$ left exposed to air, tarnishes), green (colour of malachite: $Cu_2CO_3(OH)_2$), dark blue with yellowish tint (azurite: $2CuCO_3.Cu(OH)_2$), pale red or red (native copper)'. Lead ores were stated to be (2.12.13) grayish black, like *kakamecaka* (this is the colour of galena), yellow like pigeon bile (lead ore gossan), marked with white lines (quartz or calcite gangue minerals) and smelling like raw flesh (odour of sulphur). The ores could be (2.12.14) grey like saline earth or brown like burnt earth (cassiterite: SuO_2). Iron ore was known (2.12.15) to be greasy stone, pale red colour (orange limonite: $2Fe_2O_3$, $3H_2O$) or of the colour of the *sinduvara* flower (hematite: Fe_2O_3). This matches with the characteristic of the bedded iron ores of Bihar and Orissa.

After describing the above metallic ores or *dhatus* of specific metals, the *Arthasastra* writes:

काकाण्ड भुजपत्रवर्णो वा वैकष्न्तकधातुः (2.12.16)

That which is of the colour of *kakanda* (crow's egg) or birch-leaf is *vaikrntaka* ore.

This presumably refers to the ore of *vaikrntaka* (mentioned as a metal in 2.12.23; 2.17.14 and 4.1.35), the identity of which is not known. It is due to fact that since magnetite (Fe_3O_4) is black and becomes brown on surface oxidation being changed to hematite (Fe_2O_3), *vaikrntaka* ore is magnetite. In that case *vaikrntaka* metal must be iron itself which used to be produced by the South Indians starting from the magnetite ore. It was later known as *kantaloha.*

Like *vaikrntaka* in India could be correlated with an alloy or metal made in China. This could be nickel metal or nickel alloy. It may be recalled that during the third to first century B.C.E., the Chinese copper-nickel alloy was used in Taxila and north-west India for coinage, jewellery and fancy goods. This was known as *pai-t'ung* or *tutenag(ue)* or white copper (a name used for nickel silver). It is not certain whether *vaikrntaka* metal was nickel or magnetite-based iron.

Apart from colour or external appearance, another important property of ore was known to be its density or heaviness, a variable property reflecting the metal content:

सर्वं धातुनां गौरववष्द्धौ सत्ववष्द्धिः (2.12.7)

> In the case of all ores, when there is increase in heaviness, there is increase in metal-content.

This phenomenon is due to the fact that in most cases, the metal-bearing minerals are heavier than the associated gangue minerals.

The *Arthasastra* mentioned specific uses of various metals of which gold and silver received maximum attention. There are no silver ores in India. Most of the silver is obtained as a by-product of the smelting of lead ores. The major sources of the silver and lead have been the argentiferous lead-zinc ores of the Dawdwin mines in Burma and the Zawar (24°21'N; 73° 41'E) mine, Dt. Udaipur of Rajasthan where ancient mines (earliest C-14 date obtained so far in 430 B.C.E.) are found both open cast and underground. The galena (PbS) and sphalerite (Zn,Fe)S, are microscopically distinct and capable of being separated manually, and almost certainly explains why zinc smelting developed only at Zawar. The silver content is rather low and difficult to extract. Exploration and surveying of ancient mines at Zawar reveals the major mining systems were developed in the later half of the first millennium B.C.E. Both of these are reputed to have been worthed in antiquity. It is quite likely that the silver ores were also obtained from Baluchistan and Afganistan since from early Harappan time. A small amount of silver could have come from the gold mines of South India where this metal is found naturally alloyed with gold and is believed to have been used since the Harappan period. But the impurity pattern of Mohenjodaro silver and punch-marked coins in general shows that their source was argentiferous galena rather than electrum (a natural alloy of 55%-85% gold and rest of silver) of the Indus plain. However silver plates reported from Gungeria in Dt. Balaghat of Madhya Pradesh and a punch-marked coin in the Lucknow museum has been reported with 10.53% gold means the method of separation of silver from gold was worked out during the Copper-Hoard period and not prior to it.

The duties of *suvarnadhyakṣah*, the Superintendent of Gold were defined in section 2.13 of *Arthasastra*. He was supposed to establish industrial outfits and employ *saurvarnikas* or goldsmiths, well-versedin the knowledge of not only gold and silver, but also of the alloying elements such as copper and iron and of gems which had to be set in the gold and silver wares (2.13.1-2 & 58-59). The Superintendent had to arrange security preventing smuggling, theft (2.13.30-36) and pilferage by the artisans. Various possible methods of pilferage were described (2.14.18-55).

The purity of the raw-materials—gold and silver ore, scrap and ornaments for re-use—had to be checked through streak tests on special touchstones (2.13.17-25) and chemical tests such as whitening of gold by vermilion or iron sulphate (2.13.19).

Gold-smelting was known as *suvarnapaka* (1.18.8). Gold was purified by a process of extraction with molten lead. Gold thus produced was brittle and had to be re-smelted. Thereafter, it was turned into leaves by heating, and made bright with Indus-earth when it became the base of various colours to be imparted through alloying with silver, copper, iron, etc. 92.13.6-9 & 47). Similarly, silver could also be purified using molten lead and then used as such (pure silver suitable for ornament is *tara*) or alloyed to produce different colours (2.13.10-14 & 49).

Various ornamental alloys could be prepared by mixing variable proportions of iron and copper with gold, silver and *sveta tara* or white silver which contained gold, silver and some colouring matter (2.13.50-57). Two parts of silver and one part of copper constituted *triputaka.* An alloy of equal parts of silver and iron was known as *vellaka* (2.14.20-22). This textual fact is confirmed by the gold, silver and copper objects obtained from Mauryan strata from archaeoloical sites of India and Pakistan.

The range of a goldsmith's expertise lay in setting, stringing, bead-making, making of solid and hollow objects, plating, coating, fixing and gilding (2.13.37-46; 2.14.17) *ghanam susiram samyuhyam avalepyam samghatyam vasitakam ca karukarma* (2.14.170) the main products were *kancana prsata tvastr tapantya* (2.13.33) gems set on silver and gold, heads, gold-plates and gold/silver ornaments. Setting is the fixing of geme *ksepanah kacarpanadi* (2.13.38) on gold and silver base. Stringing is weaving in threads (2.13.39). Gold plating (*tvastrkarma*) could be done on silver or copper (2.13.44-45). Gilding a silver article was done by means of the liquid or powder of sand vermilion (2.13.46).

A firm and a removable enclosing (*petaka*) was made in cases of plating, coating and fixing. Lead, copper or silver objects were coated with a gold-leaf (*acitakapatra*) on one side or with a twin-leaf fixed with lac etc. (2.14.26-33). Gold, silver or gems were embedded (*pinka*) in solid or hollow articles by pasting a thick pulp (of gold, silver or gem particles and the cementing agents such as lac, vermilion, red lead) on the object and then heating (2.14.34-42).

The qualities of 'excellent ornaments' were defined (2.13.60-61), and the goldsmiths had to make them according to the specifications, and incurring minimum losses in metals and gems, not exceeding the prescribed limits (2.14.2-10).

At the time of Mauryan empire, the natural resources of the country came to be exploited in an organized fashion. Kautilya in his *Arthasastra* describes the southern country (*Dakshinapatha*) as yielding conch-shells, *vajra* or diamonds, gem stones, pearls and gold. He further states that the southern route has many mines as *bahukhâni* (7.12). The Asokan Minor Rock Edict at Maski in Raichur district and Brahmagiri and Siddapur both in Chitradurga district of Karnataka is actually on the gold field of Kolar in Kolar district of Karnataka. The Yerragudi and Pattikonda inscriptions are in the Center of the country long famous for diamonds. The *Suvarnagiri* mentioned in the Brahmagiri edict could be identified with the golden hill of Mạski. Viewed in this light the Asokan edict of the south reflect a very material interest in the area, and though they may also proclaim the sincere Buddhism of their author they tempt us to ask whether this was not rather a case of the banner of Dharma following the prospectors than vice versa? So the emergence of different dynasties was not merely a coincidence but a historical necessity born out of several factors of which principal one must have been the regional exploitation of the natural mineral resources of the land.

The *Arthasastra* specified that the Director of Metals (*lohadhyaksa*) should establish factories for metals (other than gold and silver) viz., copper, lead, tin, *vaikrntaka*(?), *arakuta* or brass, *vrtta* (steel?), *kamsa* (bronze), *tala* (bell-metal) and *loha* (iron or simply metal?), and the corresponding metal-wares (2.12.23). Some comments on these names are called for.

In the later Vedic era, iron was known as *lohayasa* or red metal due to red colour of haematite (Fe_2O_3) and *shyamayasa* as copper, due to black colour of chalcopyrite ($CuFeS_2$). Now the question arises that if *ayas* stood for copper in Rigvedic period (*Rig Veda* VI.75.15; VIII.29.3) why it was called *lohayas* during the later Vedic era. The answer lies in the semantic changes of the word *ayas* resulting from the technological developments in the society. In early Vedic era metallurgy was in its primitive stage and varieties of baser metal such as iron was not known. In the post-*Panini* era, the nomenclature for copper changed to *tamra* and *sulva*. The element sulphur *gandhaka* was reckoned as *sulva-ari* (that which attacks *sulva* or copper) or *sulvari* which was converted to 'sulphur' in the Western Indo-European languages. Copper (75-80%) used to be alloyed with tin (25-20%) to produce *tala* or bell-metal, hence used for bells etc. and with *trapu* or tin to produce *kamsa* or bronze or the copper-tin in which tin is upto 12% and rest is copper alloy. It may be noted that tin had always been known in India as *trapu*. The other name *kastira* came into vogue later during the Greek influence in India; *kastira* was possibly derived from the Greek word *cassiterite* (tin ore). Here the tin hardens and strengthens the copper.

Another Greek word *oreichalcos*, denoting brass or the copper-zinc alloy, was probably converted to *arakuta* during the time of Kautilya. Although some copper artefacts, discovered in the archaeological sites of Lothal (22° 31'N; 72° 15'E) and Atranjikhera (27° 42'N; 78° 44'E) (2000 and 1000 B.C.E. respective dates), contained zinc, the actual manufacture of zinc in India must have strated around 400 B.C.E. in Taxila (34°12'N; 72° 56'E). The *Arthasastra's* reference to *arakuta* (2.12.23; 2.17.14; 4.1.35) and to red lead *gandhara* (2.14.35), a product from Gandhara, testify the author's familiarity with the North-West India of the period fourth century before Christian era. The author was most likely Chanakya or Visnugupta himself.

Minerals / Ores, at present, *are called physical and chemical homogenous crystal bodies formed as a result of natural physical and chemical processes. From the physical and chemical point of view* each mineral /ore represents a solid phase. The stability of minerals depends on temperature, pressure, and the chemical composition of the environment. Thus, when pressure is increased, the crystal structure of the mineral can change, while when heated the mineral is either melted or dissolved in water. Conversely, the cooling of the melt, the condensation of gases or the saturation of the solution may lead to the extraction of the solid phase. Hence, natural liquids and gases are not minerals, but constitute the environment in which minerals are either formed or changed.

From the physical point of view each mineral/ore is a dimentionally ultimate crystal body which is called a *mineral individual.* Each individual is characterized by a particular surface in the form of faces or crystals or arbitrary intergranular boundaries.

Mineral individuals are usually intergrown with other individuals, and thus form different types of rock, ores and other aggregates, the components of various geological bodies: intrusions, veins and deposits of mineral resources. They may also be found in a finely dispersed state, glassy and colloidal formations, as well as in the form of suspensions in the Earth's natural waters or in space.

In terms of their crystallochemical properties minerals are characterized by a crystal structure, i.e. by particular constituent features, such as the type of structure and the idealized chemical composition. All the mineral individuals occurring in nature are crystallochemically identical and represent a particular *mineral species.* Each mineral species has its own name, while its chemical composition is expressed by means of an idealized formula.

According to their definition minerals of the same composition with different structures pertain to different mineral species. The chemical boundaries of mineral species with variable composition and uniform structure are less determined, especially in those cases when the content of separate components undergoes considerable variation or becomes interchangeable. Minerals of the intermediate composition with stoichiometric, or multiple prime numbers, relationship of the components, differing from other minerals of the variable composition in the structural degree of order, represent independent mineral species. Thus, when the general term 'mineral' is used, it is necessary to have in mind that it has a binary designation as a synonym for the concepts of 'mineral species' and 'individual'. Minerals may be represented by their varieties, each possessing some specific features of the chemical composition and structure, as well as morphology and physical properties.

The chemical varieties are characterized by the presence of additional elements within certain natural boundaries. Sometimes they have their own names, for instance, ferrous sphalerite is called marmatite. It is dark brown to boack in colour and also known as christophite.

The structural varieties of minerals are distinguished by some specific features of the crystal structure, usually with the chemical composition remaining constant. The right-handed and the left-handed structural forms of quartz, minerals with regular and irregular location of interchangeable atoms of the elements in their structure, i.e. with various degree of order (feldspars—orthoclase and sanidine), layered minerals with various order of alternation of atomic layers in the structure, etc. can serve as an example.

The morphological varieties of minerals possess innately their own shape of individuals or mineral aggregates. A large number of them have their own names too. For instance, chalcedony—a collomorph cryptocrystalline varieties of quartz, selenite—a parallel fibrous variety of gypsum and others.

The physical varieties of minerals are characterized by specific physical properties, mainly optical ones, such as: transparence and colour. Some varieties are precious minerals or used in industry. They include the bright green beryl, or emerald, the blue aquamarine, the yellow heliodor, the transparent variety of quartz, or rock crystal, the violet amethyst, the black morion etc.

It is necessary to point out that minerals are spread not only in the earth's crust. They enter into the mantle and the deeper interior of our planet, and are scattered in the hydrosphere and the atmosphere. Minerals also constitute the Moon, numerous planets, their satellites, the asteroids, and are found in the meteorites and the smallest particles of the cosmic dust that fall on the surface of the earth.

'Artificial minerals' have synthetic analogues of minerals that are often produced in the form of large crystals. Some of them have been cultivated on an industrial scale to substitute the deficient natural raw materials: the optical quartz, diamond, corundum, emerald and others. *Technogenic minerals* also refer to the domain of mineralogical research. Their formation

in natural environment is stimulated by the productive activity of man, for instance on the walls of mine workings, external spoil heaps, in the pipes of the boreholes etc.

The crystalline products of the vital functions of various organisms are also referred to minerals. Thus, for instance, the formation of native sulphur and calcium carbonate at the expense of gypsum, is accounted for by the sulphate-reducing bacteria. Entering into the tissues of animals and plants, minerals form *organomineral aggregates*, as is the case with apatite in the bones, fluorite in the teeth, celestine and tridymite in the skeletons of radiolarians etc. On the extinction of organisms some of their minerals form accumulations of valuable mineral raw material, for instance, phosphorite, tripoli etc.

At present more than 3000 mineral species and almost the same number of varieties have been established. Every year several new minerals are discovered. The earth's crust is mainly composed of feldspar and quartz and accounts for their nearly 55% and 10% presence respectively. Pyroxenes, amphiboles, chlorites, micas, clay minerals, carbonates and others are widely spread too. A number of other mineral species are found rather often; others are of rare occurrence. Some minerals are represented by huge masses of almost monomineral rock or industrial accumulations, the deposits of mineral resources. Magnetite, magnesite, calcite, gypsum, halite and others can serve as an example. Meanwhile, there are minerals whose number is found to be inadequate for their diagnosis.

The spreading and the number of mineral species in the Earth's crust are basically determined by the spreading and the chemical properties of the atoms of the mineral-forming elements. Their ability to concentrate or to disperse in the process of the formation of minerals, i.e. their chemical activity, depends on the physical and chemical conditions of the environment. The spreading of chemical elements in the earth's crust determines the mass of predominant mineral species in terms of their number (or quantity). The multilateral character of species and varieties could be determined by the number of the possible combinations of chemical elements. The limited number of mineral species is conditioned by the fact that not all even the most spread chemical elements can interact with the formation of the stable combinations in the natural environment. The most widespread species-forming elements in minerals are: O, Si, Al, Fe, C, Na, Mg, K, Ti. Among those that are moderately spread are: S, Cl, C, Mn, H; while the number of those that are either rare or not spread at all includes B, Be, Ph, Sb, As, Se, U, Ag, Te and others. Some elements do not form independent species of minerals at all, and are in a dispersed state, as is the case with rubidium, and inert gases.

Knowledge of the mineral world, the inception and the development of mineralogical studies have always been closely connected with the practical activity of people. The growth of production forces at all the stages in the development of human society has been linked with the direct implementation of minerals or their valuable components, such as precious, non-ferrous, ferrous, and rare metals and non-metals, trace elements, and various chemical compounds.

Mineral Aggregates

Minerals are usually found in the form of irregular growths, i.e. mineral aggregates composed of one or several minerals. Mineral aggregates are distinguished by their size, form and the character of spatial distribution of mineral individuals of which they consist. There are macro-, medium- and microgranular, crystalline-granular, cryptocrystalline, equigranular and inequigranular aggregates with regular or irregular orientation of mineral grains. Mineral aggregates can be dense or massive, porous, earthy, viscous, loose, or friable. Most of the erupted, metamorphic, sedimentary rocks as well as various ores are composed of crystalline-granular aggregates of minerals. Porous and friable aggregates are typical of the products of volcanic eruptions, a large number of sedimentary rocks, soils, etc.

Mineral individuals composing incrustate aggregates in secretions can have haphazard, or rarely, regular, usually parallel orientation. Haphazardly oriented individuals are formed of numerous microcrystalline embryos with accidental fastening on the uneven walls of the voids. There are various reasons causing the orientation of crystals in secretions. The largest crystals are formed in compliance with the law of geometrical selection at the expense of the individuals whose maximum growth direction is normal to the walls in the cavity. When these individuals are fastened on an even surface of the cavity walls, the parallel oriented rod-like aggregates are usually formed.

Oriented aggregates result from the crystallization on a porous wall with the formation of parallel-fibrous aggregates of fibrous crystals, as for instance, gypsum, chalcanthite, and others. Besides, oriented effect on the individuals can be caused by the basis on which crystallization takes place. These can include the regularly oriented fissure net, the layering of enclosing rocks or a large crystal that can serve as the base during the epitaxial growth.

Various reniform, collomorphic and spherulite crusts and films as well as stalactites and stalagmites with concentrated-and-zonal structure, dendrite, branched aggregates—corallites and microcrystalline structures of aragonite and calcite are characteristic of karst caves. Analogous aggregates of carbonates and hydroxides or iron of smaller dimensions are found in cellularly porous and geode formations of the surface changes in rocks and ores. The formation of spherulite aggregates, as is the case with druse aggregates, is submitted to the law of geometrical selection. In this case each spherulite functions as a mineral individual. The formation of mineral aggregates with the growth of crystalls and spherulites, as for instance, calcite and chalcedony is also submitted to the law of geometrical selection.

Spheroidal, nodular, oval, reniform and other mineral aggregates of the rounded form are called *concretions, septarian nodules* and *oölites.* Concretions, in contrast to secretions, are formed in the process of crystallization from the centre to the periphery. They usually have radiate-fibrous or concentric-and-zonal structure. The formation of concretions is connected with collective crystallization in porous, usually sedimentary rocks at the expense of the matter dispersed in them. Oölites are of small size (up to1 cm) and are formed in mineral wells, at the bottom of the lakes in caves, in weathering crusts, in the form of pearls in mollusk shells. They sometimes form secretions with crystals.

The outer surface of concretions can be smooth and even, as is the case with phosphorite nodules and aragonite oölites, or pectinal with a clearly-cut facing, particularly characteristic of marcasite. In some cases the central part of concretions can be hollow. It is assumed that hollow oölites are formed on gaseous bubbles. Voids can also appear as a result of dissolution of recrystallization of the central part of concretions which is sometimes succeeded by the filling of other minerals. Phosphorites containing the crystals of galenite, sphalerite, calcite and other minerals can serve as an example of transformations of concretions into secretions.

Earthy, porous and *friable aggregates, blooms* and *coatings* are characteristic of zones of surface rock change or sedimentary deposit of minerals.

Inclusions and Other Heterogeneities in Minerals

Mineral individuals homogeneous in the chemical and physical sense, are extremely rare. They usually contain various inclusions of other phases, have mosaic structure and are characterized by chemical and physical heterogeneity, expressed in irregular or zonal-and-sectorial distribution of chemical or mechanical impurities, transparency or colour. In a large number of cases to establish heterogeneity microscopic investigations with the use of special methods are called for. Heterogeneous structure of minerals depends on the conditions of their formation and their subsequent change.

Protogenetic heterogeneities are of various forms and nature. Mechanical inclusions of foreign minerals intercepted in the process of mineral growth are widely spread. They form poikilitic intergrowths (from the Greek '*poikilos*'—variegated), which usually represent monocrystals with inclusions of mineral grains of enclosing rock. 'Sandy' crystals of gypsum or calcite, crystals of aegirine with inclusions of the early formed nepheline can serve as an example. Protogenetic inclusions also include the relict inclusions which are retained as a result of an incomplete replacement in various magmatic, metamorphic or metasomatic rocks. Thus, zircon of the metamorphic rocks after undergoing granitization remains in the form of rounded inclusions in the regenerated crystals of the accessory zircon of the granites.

Syngenetic heterogeneities are characteristic, for example, of drusy aggregates, crystallized gels, disintegrated solid solutions, as well as for zonal-and-sectorial crystals.

Epigenetic heterogeneities in minerals are caused by processes involving a change in their composition, depositions of newly-formed minerals in fissures or by replacing them with new minerals.

Hardened, liquid and gaseous-and-liquid inclusions in minerals are also referred to the proto-, syn- and epigenetic types of heterogeneity.

The Physical Properties of Minerals—Ores

Physical properties of minerals are found and can be measured only in the result of applying external forces and various by their nature of actions—mechanical, radiational, electromagnetic, thermal and others. The interaction with the minerals of various forms of energy is accompanied by its partial or complete absorption and dispersion, change and conversion into other forms which is expressed in the emergence of various physical effects and properties. The effect of mechanical forces on minerals is accompanied by their deformation or disintegration. The resistance of minerals to their breakdown by means of inserting a sharp object or scratching their surface is expressed through hardness.

The behaviour in the gravitational field is determined by the density of minerals. Depending on the energy of the electromagnetic field in the mineral excitation of electrons can take place accompanied by a selective absorption of light which serves as a direct cause of colour. The magnetic properties of minerals emerge in the magnetic field.

A large number of physical properties are connected with the structure and chemical composition of minerals and undergo changes with them. For other properties the presence of defects in minerals plays a significant role. In accordance with this all physical properties are subdivided into primary and secondary. The former comprise density, hardness, cleavage, optical properties, lustre, reflecting capacity and others. The primary properties have the main diagnostic value. Among the secondary ones it is necessary to bring out various semiconductive and thermoelectrical and luminescent properties, the admixing and radiational colouring, residual magnetization and others. The study of the secondary properties has the function of identifying the actual structure of minerals, the diagnosis of their genetic varieties and the information concerning the mineral-forming processes.

The importance of the physical properties of minerals is by no means confined to what has been said here. They are widely implemented in industry, science and everyday life. Thus, for the charging of the solution at drilling barite is used. Hard minerals are employed as abrasives, while the light minerals are used as concrete fillers. High transparency of quartz and fluorite in the ultraviolet, visible and infrared light determine the value of these minerals as optical raw-material; wonderful colouring in conjunction with strong lustre, transparency and a high degree of hardness distinguish numerous precious stones from other minerals.

Density

The density[10] of minerals measured in the units of mass per unit of volume (in g/cm^3) fluctuates within the wide limits and reaches 23.0 g/cm^3 (platiniridium). The more widespread minerals have density from 2.5 to 3.5 g/cm^3. A great number of the mineral species have density less than 5 g/cm^3. Among them silicates, oxygen salts and haloids are predominant. Sulphides and native metals have greater density.

Minerals can provisionally be divided into three groups: the light minerals (up to 3.0 g/cm^3), the medium (from 3.0 to 4.0 g/cm^3) and the heavy minerals (more than 4.0 g/cm^3).

The density of minerals depends on the chemical composition and the type of the structure and hence, is the most important diagnostic feature. All the minerals composed of the elements pertaining to the upper part of the periodic table[11] are light minerals. Minerals, containing heavy metals have, as a rule, greater density. Maximum density is typical of the native elements of gold, silver, and the minerals of the platinum group.

The dependence of density on the crystalline structure is particularly manifested in the polymorphic modifications of minerals and is determined by the character of the atom packing. Diamond and graphite with the density 3.5 and 2.3 g/cm^3 respectively, calcite (2.6 – 2.8 g/cm^3) and aragonite (2.9 – 3.0 g/cm^3) mentioned above can serve as an example. It is exactly the same with quartz SiO_2 with framework structure which is considerably lighter than stishovite of the same composition but having rutile-like chain structure (2.65 and 4.35 g/cm^3 respectively). In general, minerals with close packing have greater density as compared with that of minerals of chain, ribbon layered and particularly framework structure.

Several methods of determining the density of minerals are known. The most accurate ones are the pycnometric and the hydrostatic methods. The pycnometric method is based on the measurement of the mass of the mineral by means of weighing it in a pycnometer, a vessel of a given volume. The hydrostatic method presupposes the weighing of the mineral in air and in liquid.

The more unsophisticated *volumetric methods* of determining the density of minerals include the weighing with the subsequent measurement of the liquid volume displaced by the minerals. This is done with the help of the volumetric flask or special vessels-volumeters. The microscope is used to measure the volume of the mineral microsamples in the thin vessels of a given section.

The immersion methods make it possible to determine the density of minerals by comparing it with the density of heavy liquid which is determined directly by applying the Westphal balance (the hydrostatic method) or indirectly, by measuring the value of refraction ratio. The density of liquid can be regulated by heating of dilution (the thermogradient method). The method of light floats with grains attached to them is used to determine the density of heavy minerals in a liquid of lower density.

Mechanical Properties

The mechanical properties of minerals become apparent when the mechanical action, exerted on them by the external forces during compression, stretching or shock, takes place. They are expressed in the changing of their form and integrity. The following types of deformation are distinguished: elastic, brittle and plastic.

Cleavage is the ability to break down into preferred crystallographic planes with the formation of lustrous surfaces. It is the property of some minerals to break along planes related to the molecular structure of the mineral and parallel to actual or possible crystal faces. Cleavage can be manifested in one, two, three, four or six crystallographic lines. The mica crystals can be cleaved into minutest leaves. According to pinacoid its cleavage in one direction is {001}. The calcite rhombohedral crystal is easily split in free directions (along the rhombohedron), whereas in other directions cleaved planes are not formed.

If a mineral breaks along a flat plane, it is said to possess a cleavage, if it breaks with an irregular surface, it is said to be show a fracture. Thus the cleavage is a type of the tendency of a mineral to break more easily with smooth surfaces along plans of weak bounding. Hence cleavage is the property which is related to a atomic arrangement within the mineral. Because the cleavage always occurs parallel to a possible crystal face, it is designated in terms of the crystal face to which it lies parallel, such as cubic, octahedral, prismatic, basal and so on. Quartz has no cleavage at all.

The cleavage planes are parallel to the flat nets of the space lattice with a maximum interplanar distance, as for instance, in graphite. Since cleavage is connected with the crystalline structure, it functions as an important diagnostic feature of minerals.

To evaluate cleavage the following scale is used.

1. *Perfect cleavage*—when the crystal is split into very thin plates of specular surface (mica, gypsum).
2. *Good cleavage*—the crystal is split into certain directions, at any place, thus forming even, sometimes step-like surfaces; cases of irregular surfaces are extremely rare (calcite, halite, galenite).
3. *Distinct cleavage* is exemplified by the formations of both even and uneven surfaces resulting from a splitting (feldspar, hornblende).
4. *Imperfect cleavage*—even cleaved surfaces are rare. In most cases splitting results in the formation of irregular surfaces (beryl, apatite).
5. *None cleavage* occurs when the cleavage is absent and the crystals have uneven surfaces resulting from the split (quartz, cassiterite).

Among different lines crystal cleavage can be either identical or different depending on the degree of perfection.

Fracture. Minerals that have imperfect cleavage or no cleavage at all are split into uneven surfaces of the fracture. According to the character of these surfaces the fracture in minerals can be uneven (native sulphur, apatite, cassiterite), step-like (feldspars), splintery (actinolite, tremolite), shelly (quartz, chalcedony, opal), hackly (native elements—gold, brass, platinum) and other species.

Fracture means minerals which do not exhibit cleavage, break with an irregular surface. The nature of this broken surface is called "fracture". In case of fracture, the breaking should be in any other direction than the cleavage. Unlike cleavage, the fracture do not produce smooth surface. The common type of fracture are as follows:

1. Conchoidal fracture. It is a curved fracture surface showing concenteric lines like shell. Quartz and glan show conchoidal fracture.

2. Even fracture. It is a fracture surface which is almost flat. Flint shows even fracture.
3. Uneven fracture. It is a fracture surface which is irregular and rough. A large number of minerals show uneven fracture.
4. Hackly fracture. It is a fracture surface which is rough with sharp and jagged joints. Native metals show hackly fracture.

Jointing implies the ability of minerals to break down and form even surfaces on the chippings along the directions coinciding with the twinning planes, the epitaxical intergrowths, the oriented inclusions of the disintegrating products of solid solutions, the heterogeneity, and the inclusions along the zones of growth etc. For instance, jointing according to pinacoid {0001} is characteristic of corundum. It is connected with the minutest inclusions of muscovite.

Elasticity is a property shared by minerals the original size and form of which is restored after the external tension has been removed. A high degree of elasticity is characteristic of mica scale. The elasticity is connected with the strength and type of interatomic bonds. With the growth of temperature elasticity decreases. When the elastic deformation exceeds its limit, there appear brittle or plastic deformations.

Tenacity of the mineral denotes the degree or character of cohesion which is classified as follows:

1. *Sectile*: minerals which may be cut with knife but slices are not malleable.
2. *Malleable*: minerals which flatten under the hammer.
3. *Flexible*: minerals which may be bent.
4. *Elastic*: minerals which spring back after bending.
5. *Brittle*: minerals which break easily and this feature is the opposite of tough.
6. *Friable*: minerals which crumble easily.
7. *Pulverulent*: minerals which are powdery and have little or no cohesion, e.g. clay or chalk.

Brittleness is a property which manifests itself in that minerals are easily crushed as a result of the application of external forces, for instance, when the surface of the mineral is either struck by a hammer, or a knife, a steel-needle are applied to it. When such a brittle mineral as grey ore is scratched, some powder remains at the edges. It can easily be blown off. As compared with the latter, on the surface of chalcosine there remains a shining cut with even edges. Most of the native metals share exactly the same property. They change their form when they are struck and are flattened, i.e. are characterized by *malleability*.

Hardness is one of the most useful diagnostic properties of a mineral. It is defined as the resistance of a mineral to abrasion or scratching. Hardness is determining by rubbing a material of unknown hardness against one of known hardness. Mineral hardness is defined as its resistance to the mechanical disintegration which takes place under the action of a stronger body. Hardness is the most important diagnostic feature of minerals. It reflects the strength of the chemical bonds between atoms and depends on the type of chemical bonds. For minerals with an ionic type of chemical bond a low or moderate degree of hardness is characteristic, while minerals with a purely covalent bond are very hard. Halite and diamonds can be adduced as an example. Minerals with a mixed type of chemical bond usually have lowered hardness, whereas those with hydrogen and Van der Walls bonds have low hardness. In accordance with this a large number of layered, aqueous and molecular crystals are distinguished by low hardness. Minerals with a coordinate and frame types of structure are characterized by a moderately high degree of hardness. The comparison of structurally similar and chemically analogous minerals shows that with the increase of cation or anion charge, hardness increases. For instance, the hardness of calcium layered aluminosilicate-margarite $CaAl_2 [Si_2Al_2O_{10}](OH)_2$—is considerably higher than its potasiumosilicate $KAl_2[Si_3AlO_{10}](OH)_2$. In isomorphic series mineral hardness changes with the composition. Simultaneous changes in the degree of ionic-covalent chemical bond, the valency of ions and their dimensions, and consequently, the density of packing and coordination are particularly significant.

A perceptible impact on the microhardness of minerals is caused by various defects—vacancies, interstitial atoms and microimpurities, dislocations and the degree of perfection of the crystalline individuals. The theoretical stability of the crystals is known to be considerably greater than the actual stability. Only the undislocational filamentary crystals (Vickers) have a stability corresponding to the one that has been calculated.

There are several methods of determining hardness. In mineralogical practice *Mohs' scale*[12] is employed. The following minerals listed according to the degree of hardness can serve as standards of Mohs' scale of hardness:

1. Talc $Mg_3[Si_4O_{10}]OH_2$
2. Gypsum $Ca[SO_4] \cdot 2H_2O$
3. Calcite $Ca[CO_3]$
4. Fluorite CaF_2
5. Apatite $Ca_4[PO_4]_3(F,Cl)$
6. Feldspar (orthoclase) $K[Si_3AlO_8]$
7. Quartz SiO_2
8. Topaz $Al_2[SiO_4](F,OH)_2$
9. Corundum Al_2O_3
10. Diamond C

Mineral hardness is usually determined by the scratching of one mineral against another. If, for instance, apatite (hardness 5) scratches the mineral under analysis, but fluorite (hardness 4) does not, hence, the hardness of the mineral under investigation is to be approximately 4.5 In absence of hardness testing minerals, the following material may be used to determine approximate hardness: (i) A fingernail will scratch upto about 2.5 Mohs scale (i.e. not calcite), (ii) A window glass will scratch up to about 5 Mohs scale (i.e. not felspar), (iii) A penkife will scratch upto about 6.5 Mohs scale (i.e. not quartz).

It is necessary to point out the relative character of Mohs' scale. If the hardness of talc is 1, and of gypsum 2, it does not mean that gypsum is twice as hard as talc. The same may be said of other standard mineral.

Colour of mineral-ores. The colour of minerals is the most obvious and conspicuous outer property of a large number of minerals. Minerals are distinguished by an extraordinary variety of colour and shades, richness and intensity. Some mineral species are characterized by a constant colour, which enables to determine them almost unerroneously. It would suffice to adduce vermilion, golden yellow pyrite, geeen malachite, blue lazurite. Other minerals, such as tourmaline, beryl, garnet, fluorite, calcite may be of various colour. Thus, for instance, the crystals of tourmaline are found to be black, green, blue, brown, rose or even lacking in any colour at all.

In a large number of minerals the colour is spread evenly. However, multicoloured (polychromatic) minerals with mottled or regular zonal or sectorial distribution of different colour are also found. For instance, sometimes, one end of the tourmaline crystals is black, while the other is green, the middle part remaining colourless or rose. Its green variety called *indiclolite* occurs at Hazaribagh (Jharkhand), and in the Padar district of Jammu and Kashmir.

Fluorite, quartz, calcite, garnet, topaz and others are also often found in the form of *ploychromatic* crystals.

The intensity with which colour manifests itself depends on the aggregate condition and the character of the surface of minerals. In large crystals and coarse-grained aggregates the colour of many minerals is usually darker than in smaller grains, finely-granulated or dispersed aggregates. Thus, the outwardly black hematite (Fe_2O_3) in earthy aggregates is dark-red, while the almost black aegirine ($NaFe(SiO_3)_2$)in finely-shaped aggregates has a light-green colour. On the other hand the golden-yellow pyrite (FeS_2) in earthy aggregates is black. As a rule, the transparent crystals of minerals have very little colour or are achromatic (colourless) at all. In dispersed powdery, porous, earthy and fine-grained aggregates or in crystals containing many inclusions the dark minerals are usually brighter.

The measure of intensity of absorption or the light transmission passing through the crystal is *the transmission* $T = I/I_o$ where I_o is the intensity of the initial luminous flux; *I*—the intensity of the latter when it has passed through the crystal; and the *optical density* $D = \lg 1/T$. In accordance with the Bouguer-Lambert law

$$I = I_0 e^{-\acute{a}t}$$

where *t* is the thickness of the crystal (cm); *á* is the coefficient of absorption, which characterizes the nature of the mineral.

Fluorescence. Some mineral-ores when exposed in the sunlight or ultraviolet light, produce a colour quite different from their dose. Thus green or colourless florite shows a blue or purple colour in ultraviolet light. The property of the minerals is called fluorescence are seen in calcite ($CaCO_3$)and scheelite ($CaWO_4$). Fluorescence[13] differ from phosphorescence[14] in that fluorescence ceases within about a microsecond of stimulus ceasing whereas phosphorescence can persist for minutes. The examples of minerals which show phsophorescence are diamond and sphalerite (Zn,Fe)S..

Thermal Properties

When minerals are heated the atoms entering into their composition are shifted in respect to the state of equilibrium. These shifts are mainly of fluctuating character of particular frequency. With the rise of temperature the amplitude of fluctuations is increased. The increase in the distance between the fluctuating atoms leads to changes of repulsive and attractive forces. Consequently, the kinetic energy is transformed into the potential energy, and vice versa. The equilibrium (theoretical) state in the structure corresponds to the minimal value of the potential energy for each atom (or atoms of one type). The potential energy of the fluctuating atom increases with the temperature. Variations of the vibrational energy also depends on the amplitude of vibration (i.e. on the value of the shift of atoms in respect to the equilibrium state), on the mass of atoms, the coordination number of atoms and other factors.

Thermal capacity of minerals C_v determines the amount of energy required to heat 1 mole by 1^0. At absolute zero (K) $C_v = 0$ at higher temperatures it increases proportionally to T^3, and then, beginning with a certain temperature (the Debye temperature) it becomes constant.

Thermal expansion is mainly determined by the growth of amplitude of fluctuating atoms when the heating of minerals takes place. Accordingly, the distance between the atoms increases, and the chemical bond becomes weaker. *The coefficient of linear expansion* is an anisotropic value and depends on the crystalline structure of minerals. Thermal expansion is also connected with a shift of atoms from the state of equilibrium and the transition to interstitial positions, i.e. with the formation of thermal defects—anion vacancies and interstitial atoms. The concentration of thermal defects is in an exponential dependence on temperature. Hence, a greater degree of thermal expansion corresponds to a greater concentration of vacancies in the structure of minerals. For example, metals with vacancies increase in volume when heated to the melting point by a tenth of a percent.

The melting of minerals takes place when the distance between the atoms increases and there is an increase in the specific volume of the mineral which is called *unconfined space*. Practically all the minerals melt with an increase in unconfined space by 2-6%. Ice and bismuth, whose structure is more dense in the liquid state, can be referred to as exceptions. *The melting heat* is the energy required to destroy the crystalline structure with the formation of the melt at a certain temperature immanent to each mineral, or *the melting point*. The latter decreases when the concentration of point defects and isomorphic impurities becomes greater. If melting takes place directly, by means of a transition into the liquid state with the retention of the composition, it is called *congruent*. Congruent melting is characteristic of metals simple in their composition, salts, oxides etc. In those cases when intermediate phases differing by the composition from the original mineral and melt are formed in the process of melting, the melting is called *incongruent*. The intermediate phases are usually formed during the melting of silicates, salts etc. whose composition is complex.

The temperature and the type of melting are accurately determined by experimental means presupposing absorption or emission of heat at uniform heating. Retardation in the rise of temperature at heating is connected with the processes of the absorption of melting heat. The heating of minerals is usually accompanied by the diffusive shifts of atoms. Under these circumstances in some of them disordering takes place. In others there is a homogenization of phases (the intergrowths of the products of the disintegration of solid solutions). In some of them there are phasal transformations – polymorphic transitions that are often accompanied by an increase in the volume, for example at the polymorphic transformation of aragonite → calcite.

Thermal decomposition of minerals takes place when the temperature is below that of the melting point and is expressed in *thermal dissociation* of, for example, calcite $Ca[CO_3] \rightarrow CaO+CO_2 \uparrow$ or hematite $Fe_2O_3 \rightarrow Fe^{2+} Fe^{3+} O^{+2} + O_2 \uparrow$, and *thermal dehydration* of minerals, containing water, for example, gypsum $Ca[SO_4].2H_2O \rightarrow Ca[SO_4] + 2H_2O \uparrow$, when heated in the open air, i.e. in the oxidizing atmosphere, some minerals are oxidized or burned. When sulphides, oxides and salts are heated, the metal bead is produced in the conditions of recovery.

Sublimation or vaporization, of minerals when they are heated is found to be typical of native mercury, cinnabar, realgar, orpiment, as well as graphite, bismuth, gold and other metals, particularly in the restoring conditions or in vacuum. The heating of metamictic minerals is accompanied by the liberation of energy of crystallization, leading to their excessive overheating and a bright glow, which becomes particularly noticeable in pyrochlore, samarskite etc.

It is also necessary to lay particular stress on the influence of heating on the colour of minerals. Under heating some colourless minerals of iron (siderite) darken, which is connected with oxidation Fe^{2+}.

Minerals with radiation colour usually change their colour or become colourless under heating. Thus, for instance, the radiation colouring of the green spodumene, when heated to 250° C becomes purple; further heating of spodumene up to 500° C makes the mineral lose its colour altogether.

The study of thermal properties and phenomena, which, in the process of heating, are accompanied by absorption or emission of heat, as well as by a loss or gain of mass, is carried out by employing the methods of thermal analysis with the help of special instruments—pyrometers, thermographs, thermoscales and more complex apparatuses. Microcalorimeters are used to measure the exact thermal constants—the heat of melting, dissolution *et al.*

The study of thermal properties of minerals and the phenomena pertaining to them is of great significance in the diagnostics of minerals, when their content and form of finding water, the valency of ions, in the measuring of the temperature of phasal transformations, dissociations, dehydration and melting takes place. The acquisition of thermal constants is necessary to interpret the reactions of native mineral formation.

The knowledge of the thermal stability of minerals is also important so as to ascertain the conditions of their formation and it is particularly important in the technological processing of mineral raw materials.

The absorptional ability of minerals is varied in its nature and is manifested in that minerals absorb on their surface molecules and ions. The absorptional ability is particularly great with the dispersed minerals due to a large density surface, measured in the units of area by the units of mass (cm^2 /g). Absorption is said to be chemical if strong chemical bonds are formed between the absorbent and the absorbate. Physical absorption is the other type of absorption, which is characterized by the fact that the absorbent molecules can be easily scaped. In the first case absorption can be likened as to the formation of chemical composition, containing combined atoms, belonging to both the absorbent and the absorbate simultaneously. In the second case these compositions are not formed. Hygroscopicity, i.e. the absorption of moisture from the air by the insoluble minerals can be referred to as an example of physical absorption of water (kaolinite, halloysite, and other clay minerals). These minerals stick to the tongue or the lips. As distinct from the latter some easily soluble minerals are hygroscopic to such extent that become dissolved (carnallite, bishofite and others).

The wettability of minerals wholly depends on the crystallochemical characteristics and type of non-compensated chemical bond typical of the surface of chippings. Hydrophilic minerals are easily wetted in water as compared with the hydrophobic ones, that are not wetted and are driven out as to the borderline separating the liquid from the air. This property of the minerals is used when the floatation enrichment of ores takes place. Mineral particles with natural floatability—hydrophonic ore minerals—become adhesive to air-bubbles, come on the surface and concentrate in the foam, which together with the ore minerals is mechanically separated from the hydrophilic minerals. Hydrophobic minerals include metals with fully compensated charges (gold, copper, platinum and others), layered minerals with a markedly perfect cleavage and the molecular crystals with weak hydrogen bonds or vandar Walls bonds (graphite, talc, molybdenite, sulphur, realgar and others). Weak hydrophobic minerals include minerals-semiconductors with a complete or partially compensated covalent or covalent-metallic bonds (sulphides—galena, sphalerite, pyrite, chalcopyrite and others, as well as diamond). Natural floatability is not characteristic of hydrophilic minerals—dielectrics with uncompensated strong ionic or ionic-covalent bonds (barite, calcite, quartz, scheelite, feldspar, spodumene etc.).

Common Minerals

Silicate minerals are by far the most common rockforming minerals containing silicon, oxygen and one or more metals. They constitute about 95% of the earth's crust. The important silicate minerals are felspars, felspathoids, quartz, pyroxenes, amphiboles, micas and olivines. The other minerals which are described here are garnets, chlorite, serpentine, aluminium silicates, calcium minerals and some ore minerals.

Felspar Group[15]

A group of rockforming minerals that make up about 6% of the earth crust. Their general formula is MAl $(AlSi)_3 0_8$, where M can be K, Na, Ca, Ba, Fe. The felspars are the most abundant of all minerals. They fall into two main series: (I) K – Na felspars, called the "alkali felspars", and (ii) Ca – Na felspars, called the "plagioclase felspars".

All felspars show closely related physical properties. Cleavage – good in two directions at an angle of 90^0 (approximately). H – 6. Sp.gr – 2.55 – 2.63. Lustre – Vitreous. Colour – white, grey, pink or green. Habit – generally tubular. Use – in the manufacture of porcelain.

K-Felspars. The important members of this group are "orthoclase" and "microcline". Each of these has the composition $KAlSi_3O_8$ but orthoclase crystallizes in the orthorhombic system and microcline in the triclinic system. Potash felspars alter readily into clay minerals, especially "kaolinite" $(Al_2Si_2O_5(OH)_4)$.

Plagioclase Felspars. Few plagioclase felspars form a completesolution series from pure "albite", $NaAlSi_3O_8$ to pure "anorthite", $CaAl_2Si_2O_8$. The important intermediate members are "oligoclase", "andesine", "labradorite" and "bytownite".The plagioclase felspars crystallize in the triclinic system and are distinguished from K-felspars by the presence of repeated albite twin striations visible on cleavage surfaces. The properties of the various members of the plagioclase series vary in a uniform manner, with the change in the chemical composition. The plagioclase felspars usually alter to clay minerals or "sericite" (fine grained micaceous material).

Occurrence. (i) The alkali felspars which include orthoclase, microcline and albite, have a similar occurrence. The type of felspar present is the important basis of the classification if igneous rock. They are commonly found in igneous rocks, such as granites, pegmatites, syenites, rhyolites and trachytes. (ii) Oligoclase is commonly found in granodiorites and monzonites. (iii) Labradorite is a common constituent of gabbros, basalts and anorthosites.

Felspathoid Group

The felspathoids are anhydrous framework silicates. They contain less silica than felspars. The important members of this group are nephaline, leucite and sodalite.

Nephaline : (Na, K) $AlSiO_4$

Crystal system – hexagonal. Cleavage - poor. *H* - 6. Sp.gr. – 2.60 – 2.63. Lustre – vitreous to greasy. Colour – colourless, white, grey or yellowish. Habit – massive. Occurrence – mostly in nephaline syenites. Use – in the manufacture of glass and ceramics.

Leucite : $KalSi_2O_6$

Crystal system – cubic (above 625^0 C) and tetragonal (below 625^0 C). Cleavage – very poor. *H* - 5½ - 6, Sp.gr. – 2.47 – 2.50. Lustre – vitreous. Colour – white to grey. Habit – trapezohedral crystals. Occurrence – In volcanic rocks only.

Sodalite : $Na_8 (AlSiO_4)_6Cl_2$

Crystal system – cubic. Cleavage – poor. *H* - 5½-6, Sp.gr. – 2.15-2.3. Lustre – vitreous. Colour – usually blue, also white, grey, green. Habit – commonly massive. Occurrence – with nephaline in nephaline syenites.

Silica Group[16]

Quartz : SiO_2

Crystal system - hexagonal. Cleavage – absent. *H* - 7. Sp.gr.- 2.65. Fracture – conchoidal. Lustre – vitreous. Colour – colourless, white or with a wide range of tints. Habit – prismatic crystals or massive.

The important varieties of quartz are: (i) "Rock crystal" – colourless quartz, commonly in distinct crystals. (ii) "Amethyst" – Transparent quartzwith purple colour. (iii) "Smoky quartz" – colour in shades of grey or brown. (iv) "Rose quartz" – Colour – a red rose or pink. (v) "Milky quartz" – Colour – milky white. Cryptocrystalline forms of quartz are: (a) "Chalcedony" – Waxy lustre and botryoidal form. (b) "Agate" – A banded variety with layers of chalcedony having different colours. (c) "Jasper" – Red chalcedony stained by hematite. (d) "Flint" – Dark grey siliceous nodules. (e) "Chert" – Light coloured massive quartz. (f) "Opal" – An amorphous variety of quartz ($SiO_2.nH_2O$)

Occurrence. Quartz occurs in most igneous, metamorphic and sedimentary rocks. Some sandstones and quartzites are composed almost entirely of quartz. It is also found abundantly as a gangue mineral in mineral veins. Agate occurs in volcanic lavas as cavity filling.

Pyroxene Group[17]

Pyroxenes form an important group of rockforming minerals. They have a general formula $X_2Si_2O_6$ in which X is usually Mg, Fe, Al, Ca or Na. Some aluminium may partly replace silicon. The common pyroxenes are Ca, Mg and Fe silicates. They are characterized by two cleavages which intersect almost at right angles. Pyroxenes are broadly divided into two groups: (i) ortho-pyroxenes, and (ii) clino-pyroxenes.

Orthopyroxenes. These pyroxenes crystallize in the orthorhombic system and contain very little calcium. The "*enstatite*" ($MgSiO_3$) and "*hypersthene*" [$(Mg,Fe)SiO_3$)] belong to this group.

Clinopyroxenes. These pyroxenes crystallize in the monoclinic system and contain either calcium or Na, Al, Fe (ferric) or Li. "*Diopside*", "*hedenbergite*" and "*augite*" are the important members of this group.

Enstatite : $MgSiO_3$

Hypersthene: $(Mg,Fe)SiO_3$

Crystal System – orthorbombic. Cleavage – good. *H* - 5½-6. Sp.gr. 3.2-3.6. Lustre – vitreous. Colour – greyish, yellowish or greenish. Habit – massive or lamellar.

Occurrence. Magnesium rich orthopyroxenes are a common constituents or peridotites, gabbros, norites and basalts. They are found associated with augite, olivine and plagioclase. The orthopyroxenes alter to serpentinous products or fibrous amphibole.

Diopside : $CaMgSi_2O_6$

Hedenbergite : $CaFeSi_2O_6$

Augite : $(Ca,Na)(Mg,Fe,Al)(Al,Si)_2O_6$

Crystal System - monoclinic. Cleavage – at 87^0 and 93^0, imperfect, *H* – 5-6, Sp. gr. – 3.2-3.3. Lustre – vitreous. Colour – white to light green in diopside, deepens with increase of iron. Augite is black. Habit – crystals show square or eight-sided cross-section. Also granular, massive and lamellar.

Occurrence. Diopside and hedenbergite occur in metamorphic rocks. Augite occurs in gabbros, basalts, diorites and some syenites.

Amphibole Group[18]

The amphibole group includes a number of minerals which have a closely related structure and chemical composition. The amphiboles contain hydroxyl groups in the structure. The angle between two sets of cleavages is 124^0 and is characteristic of amphiboles. "*Tremolite*" (hydrous silicate of Ca and Mg), "*actinolite*" (hydrous silicate of Ca, Mg and Fe) and "*hornblende*" (hydrous silicate of Na, Ca, Fe and Al) are the important members of amphibole group.

Tremolite : $Ca_2Mg_5Si_9O_{22}(OH)_2$

Actinolite : $Ca_2(Mg,Fe)_5Si_8O_{22}(OH)_2$

Crystal System—monoclinic. Cleavage – perfect. *H* – 5-6, Sp.gr. – 3.0-3.3. Lustre – vitreous or silky. Colour – white or grey in tremolite and green in actinolite. Habit – termolite is often bladed and sometimes in silky fibres. Actinolite is often in radiating aggregate. Also fibrous.

Occurrence. These minerals are commonly found in metamorphic rocks. Tremolite occurs in metamorphosed dolomitic limestones and actinolite is a characteristic mineral of the greenschish facies of metamorphism.

Uses. The fibrous varieties are used as asbestos.

Hornblende : $(Ca,Na)_{2-3}(Mg,Fe,Al)_5Si_6(SiAl)_2O_{22}(OH)_2$

Crystal Sstem – monoclinic. Cleavage – good prismatic. *H* – 5-6. Sp.gr. – 3.0-3.4. Lustre – vitrous. Colour – dark green to black. Habit – crystals with six sided cross-section, also massive or granular.

Occurrence. Hornblede is an important rockforming mineral. It occurs commonly in syenites and diorites. It is a major mineral of amphibolites.

Mica Group[19]

There are two common varieties of mica: (i) "*muscovite*" which is rich in aluminium, and (ii) "*biotite*" which is rich in iron and magnesium. The physical properties of both the micas are similar. They crystallize in the monoclinic system and have one perfect basal cleavage. They occur infoliated form and the mineral can be split into thin elastic sheets.

Muscovite : $KAl_2(AlSi_3O_{16})(OH)_2$

Crystal System – monoclinic. Cleavage – perfect basal. *H* – 2-½. Sp.gr.- 2.76-2.88. Lustre – pearly or silky. Colour – colourless, pale, shade of green and brown. Transparent in thin sheets. Habit – usually tubular crystals, foliated.

Occurrence. Muscovite is a characteristic mineral of granites and granite-pegmatite. In pegmatites large crystals of muscovite are found. Muscovite is also common in schists and gneisses.

Biotite : $K(Mg,Fe)_3(AlSi_3O_{16})(OH)_2$

Crystal System – monoclinic. Cleavage – perfect basal, *H* - 2½-3. Spgr. – 2.8-3.2. Lustre – pearly or splendent. Colour – dark green, brown to black. Habit – usually in irregular foliated masses.

Occurrence. Biotite is found in granites, syenites, mica-lamprophyres, gneisses and schists.

Olivine Group[20]

This group includes a series of minerals that crystallize in the orthorhombic system. These minerals range in composition from "*forsterite*" Mg_2SiO_4 to "*fayalite*", Fe_2SiO_4. The composition of common olivine is $(Mg.Fe)_2SiO_4$

Olivine : $(Mg.Fe)_2SiO_4$

Crystal System – orthorhombic. Fracture – conchoidal. *H* - 6½-7. Sp.gr.- 3.27-4.37. Lustre – vitreous. Colour – pale, yellow green or olive green. Habit – granular.

Occurrence. Olivine occurs mainly in the dark coloured igneous rocks such as gabbro, peridotite and basalt. The rock, dunite is made up almost wholly of olivine. Olivine alters very readily to serpentine.

Garnet Group[21]

The garnets comprise a group if isomorphus minerals with the general formula $X_3Y_2(SiO_4)_3$, where X may be Ca, Mg, Mn or Fe^{2+} and Y may be Al, Cr^{3+} or Fe^{3+}. The common garnets are: (I) "*Pyrope*", $Mg_3Al_2Si_3O_{12}$, (ii) "*Almandite*", $Fe_3Al_2Si_3O_{12}$, (iii) "*Spessarite*", $Mn_3Al_2Si_3O_{12}$ and (iv) "*Grossularite*", $Ca_3Al_2Si_3O_{12}$.

Crystal System. – cubic. Cleavage – absent. Fracture – uneven. *H* - 6½-7½. Sp.gr. – 3.5-4.3. Lustre – vitreous to resinous. Colour – varying with composition. Most commonly red, also brown, yellow or green. Streak – white. Habit – dodecahedral crystals common, also granular.

Occurrence. Garnets occur abundantly in some metamorphic rocks such as mica-schists, hornblende-schists, and gneisses. They also occur as an accessory mineral in some igneous rocks.

Chlorite Group[22]

The chlorite group includes a number of minerals which have similar physical properties. The important mineral of this group is "*chlorite*".

Chlorite : $(Mg,Fe,Al)_6(Al,Si)_4O_{16}(OH)_8$

Crystal System – monoclinic. Cleavage – perfect basal. *H* – 2-2½. Sp.gr. – 2.6-3.3. Lustre – vitreous, resinous or dull. Colour – green of various shades. Habit – usually foliated masses.

Occurrence. Chlorite is a common mineral in low grade metamorphic rocks, such as chlorite-schists. In igneous rocks it occurs as an atteration product of pyroxenes, amphiboles and biotite.

Serpentine Group[23]

Serpentine occurs in two distinct forms: (i) "*Antigorite*", a platy variety, and (ii) "*Chrysotile*", a fibrous variety.

Antigorite and Chrysotile : $Mg_3Si_2O_5(OH)_4$

Crystal System – monoclinic. Cleavage – basal in platy variety and none in fibrous crysotile. *H* – Variable 3-4. Sp.gr. 2.5-2.6. Lustre – greasy in the massive varieties, silky in fibrous varieties. Colour – various shades of green, also brownish. Habit – lamellar, platy or fibrous (chrysotile).

Occurrence. Serpentine is usually formed by the alteration of magnesium silicates such as olivine, pyroxene and amphibole. It is found associated with magnesite, chromite and magnetite. Serpentinite is a rock which is made up mostly of the variety antigorite. It is formed by the alteration of olivine bearing rocks.

Clay Mineral Group[24]

Clays consist of mainly of a group of crystalline substances known as the "*clay minerals*". They are all essentially hydrous aluminium silicates.

Kaolinite : $Al_2Si_2O_5(OH)_4$

Crystal System – triclinic. Cleavage – perfect basal. *H* – 2. Sp.gr.- 2.6. Lustre – dull earthy. Colour – white, sometimes brown or grey. Streak – white. Habit – clay like masses.

Occurrence. Kaolinite is a secondary mineral formed by the alteration of alkali felspars. It is the chief constituent of clay.

Talc : $Mg_3Si_4O_{10}(OH)_2$

Crystal System – monoclinic. Cleavage – perfect basal. *H* – 1. Sp.gr.– 2.7 – 2.8. Lustre – pearly or greasy. Colour – white, grey, green, or dark brown. Streak – white. Habit – usually foliated masses. The compact and massive variety is known as "steatite" or "soapstone". Feel – greasy.

Occurrence. Talc is a secondary mineral formed as a result of the alteration of olivine, pyroxene and amphibole. It is often derived from ultrabasic igneous rocks. It is also found in some schists in association with actinolite.

Al_2SiO_5 Group

The three important members of Al_2SiO_5 group are (i) "*Andalusite*", (ii) "*Sillimanite*", and (iii) "*Kyanite*". These minerals occur in metamorphosed aluminous rocks, such as pelitic schists.

Andalusite : Al_2SiO_5

Crystal System – orthorhombic. Cleavage – poor. Fracture – uneven. *H* - 7½. Sp.gr. – 3.16-3.20. Lustre – vitreous. Colour – flesh red, reddish brown or live green. Habit – prismatic crystals.

Occurrence. Andalusite occurs typically in thermally metamorphosed argillaceous schists and in regionally metamorphosed rocks formed under low pressure conditions.

Sillimanite : Al_2SiO_5

Crystal System – orthorhombic. Cleavage – perfect (010). *H* – 6-7. Sp.gr.- 3.23. Lustre – vitreous. Colour – brown, pale, green or white. Habit – long slender crystals, often in parallel groups, frequently fibrous.

Occurrence. Sillimanite occurs in schists and gneisses of high grade regional metamorphism.

Kyanite : Al_2SiO_5

Crystal System – triclinic. Cleavage – perfect (100). *HCl* – 5 along length of crystals and 7 at right angles to this direction. Sp.gr.- 3.55 – 3.66. Lustre – vitreous to pearly. Colour – usually blue to white, also grey or green. Habit – crystals usually flat, bladed, also in bladed aggregates.

Occurrence. Kyanite occurs in regional metamorphic schists and gneisses together with garnet, staurolite, mica and quartz.

Miscellaneous Minerals

Beryl : $Be_3Al_2Si_6O_{18}$

Crystal System – hexagonal. Cleavage – poor. Fracture – conchoidal to uneven. *H* – 8. Sp.gr. 2.6 – 2.8. Lustre – vitreous. Colour – pale green, bright green, bluish green, yellow or white. Habit – hexagonal prisms.

Occurrence. The crystals of beryl occurs in granites, pegmatics, biotite-schists, gneisses and in pneumatolytic hydrothermal veins. Beryl crystals in some pegmatites grow to very large sizes.

Staurolite : $Fe_2Al_2O_6(SiO_4)_4(O, OH)$

Crystal System – monoclinic. Cleavage – distinct. Fracture – uneven. *H* – 7-7½. Sp.gr. 3.65-3.75. Lustre – resinous to vitreous. Colour – red – brown to brownish black. Habit – prismatic crystals.

Occurrence. Staaurolite is formed during regional metamorphism of aluminium rich rocks. It occurs commonly as porphyroblasts in schists and gneisses, often in association with garnet, kyanite and mica.

Tourmaline : $Na(Mg,Fe)_3Al_6(BO_3)_3Si_6O_{18}(OH)_4$

Crystal System – hexagonal. Cleavage – very poor. Fracture – uneven. *H* – 7-7½. Sp.gr. 3.0-3.25. Lustre – vitreous to resinous. Colour – usually black, also brown, dark blue, colourless, pink, green and blue. Habit – prismatic crystals often with triangular cross-section, parallel or radiating crystal groups common.

Occurrence. Tourmaline occurs commonly in granite –pegmatites or in some granites which have undergone metasomatism by born rich fluids.

Topaz : $4[Al_2SiO_4(F, OH)_2]$

Crystal System – orthorhombic. Cleavage – perfect basal. *H* – 8. Sp.gr. 3.4-3.6. Lustre – vitreous. Colour – colourless, yellow, pink, bluish, greenish. Habit – prismatic crystals, also granular.

Occurrence. Topaz is found in cavities in granites, granite-pegmatites and high temperature veins, and in a few rhyolites. Topaz is usually associated with cassiterite, tourmaline, fluorite and muscovite.

Calcium Minerals

Calcite : $CaCO_3$

C*rystal System* – hexagonal. Cleavage – perfect. *H*-3. Sp.gr.- 2.71. Lustre – vitreous. Colour – usually white to colourless. Also grey, red, green, blue, yellow or brown. Habit – four common forms are tabular, prismatic, rhombohedral and scalenohedral. Besides these fibrous, granular and stalactitic forms are also found.

Occurrence. The origin of calcite may be igneous as in intruisive carbonatites, sedimentary, such as limestone and metamorphic as in marbles. Calcite is commonly found in metallic ore veins and calcite veins are also found in various rocks.

Aragonite : $CaCO_3$

Crystal System – orthorhombic. Cleavage – distinct. Lustre – vitreous. *H* - 3½-4. Sp.gr.- 2.95. Colour – colourless, white, pale yellow, grey, rarely green or violet. Habit – prismatic crystals, also in tubular or stalactitic forms.

Occurrence. Aragonite is much less common than calcite. It is the main component of shells of many organisms. Aragonite occurs in some sedimentary deposits.

Dolomite : $CaMg(CO_3)_2$

Crystal System – hexagonal. Cleavage – perfect rhombohedral. *H* - 3½-4. Sp.gr.- 2.85. Lustre – vitreous to pearly. Colour – white, yellow, brown, pinkish or colourless. Habit – crystals show curved faces. Also granular and massive.

Occurrence. Dolomite occurs as extensive sedimentary beds. It is generally thought to be formed by the replacement of calcium carbonate of limestone by dolomite.

Gypsum : $CaSO_4.2H_2O$

Crystal System – monoclinic. Cleavage – perfect. *H* – 2. Sp.gr.- 2.32. Lustre – usually vitreous. Also pearly and silky. Colour – colourless, white, grey, sometimes yellow, red or brown. Habit – tabular or prismatic crystals. Also fibrous, foliated granular and massive. Varieties – "*Satinspar*" is a fibrous gypsum with silky lustre. The fine grained massive variety is called "*alabaster*". A variety that yields broad colourless and transparent cleavage folia, is called "*selenite*".

Occurrence. Gypsum occurs as an evaporite deposit. It is also found as a secondary mineral, deposited from percolating groundwater.

Anhydrite : $CaSO_4$

Crystal System – orthorhombic. Cleavage – perfect. *H* – 3-3½. Sp.gr.- 2.89-2.98. Lustre – vitreous to pearly. Colour – white, colourless, grey and sometimes bluish or red. Habit – usually massive, fibrous or grandular. Concretionary forms are also found. *Aliteration* - by hydration anhydrite changes to gypsum with an increase in volume.

Occurrence. Anhydrite is found in bedded evaporite deposits. Most of the deposits of anhydrite are thought to be secondary after gypsum.

Fluorite : CaF_2

Crystal System – cubic. Cleavage – perfect octahedral. *H* – 4. Sp.gr.- 3.18. Lustre – vitreous. Colour – light green, yellow, bluish green or purple. Also colourless white or brown. Habit – usually cubic crystals. Also massive or granular.

Occurrence. Commonly found in hydrothermal veins. It is found associated with galena and sphalerite. Fluorite also occurs in pneumatolytic tin veins and in some granites.

Apatite : $Ca_5(PO_4)_3(F,Cl,OH)$

Crystal System – hexagonal. Cleavage – poor. *H* – 5. Sp.gr. – 3.15-3.20. Lustre – vitreous to resinous. Colour – some shade of green or brown. Also blue, violet or colourless. Habit – long prismatic crystals, tabular. Also massive or granular.

Occurrence. Commonly found in all igneous rocks including pegmatites and hydrothermal veins. Apatite is also found in contact and regionally metamorphosed rocks and in sedimentary rocks.

Aluminium Minerals

Bauxite: (A mixture of hydrous aluminium oxides)

Crystal System – bauxite is a mixture of several minerals, such as "gibbsite", "*boehmite*" and "*diaspore*". Cleavage – none. Fracture – uneven. *H*

– 3-2.5. Lustre – dull earthy. Colour – white, gray, yellow or brown. Streak – yellow to brown. Habit – pisolitic. Also massive.

Occurrence. Bauxite is a secondary product, commonly formed under subtropical to tropical climatic conditions by the leaching away of silica from aluminium-bearing rocks.

Corundum : Al_2O_3

Crystal System – hexagonal. Cleavage – absent. Fracture – uneven. *H*- 9. Sp.gr. – 4.02. Lustre – vitreous to adamantine. Colour – commonly Muish grey to brown. Gem varieties are transparent to translucent. "Ruby" is of red colour and "*sapphire*" is of blue of colour. Habit – commonly tabular crystals, also massive or granular (entry).

Occurrence. Corundum is found in some metamorphic rocks, such as mica-schist and gneiss. It is also found in pegmatites and some nephaline syenites.Gem stones are found in placer deposits.

Iron Minerals[25]

Crystal System – cubic. Cleavage – absent. *H* – 6. Sp.gr.- 5.18. Lustre – metallic to dull. Colour – iron-black. Streak – black. Special character – strongly magnetic. A natural magnet is called "*iodestone*". Habit – octahedral crystals, massive or granular.

Occurrence. Magnetic is a high temperature mineral. It occurs as an accessory mineral in igneous rocks, in contact and regionally metamorphosed rocks, and in high temperature sulfide veins.

Hematite : Fe_2O_3

Crystal System – hexagonal. Cleavage – absent. *H* – 5-6. Sp.gr.- 5.26. Lustre – metallic to dull. Colour – reddish brown to black. Streak – dark red. Habit – crystals thin tabular, usually massive. Sometimes in betryoidal forms. It may also be micaceous and foliated. Varieties --red earthy variety is called "*red ocher*". Platy and metallic variety is known as "specularite".

Occurrence. The hematite is found in thick beds of sedimentary origin. It is also found in metamorphosed sediments and contact metamorphic deposits.

Goethite : FeO (OH)

Crystal System – orthorhombic. Cleavage – perfect. *H* – 5-5½. Sp.gr. 3.3-4.3. Lustre – adamantine to dull. Silky in fibrous varieties. Colour – dark brown to yellowish brown. Streak – yellowish brown. Habit – elongated prismatic crystals, tabular, massive, stalactitic or radiating fibrous.

Occurrence. Goethic is a secondary mineral. It is found in the oxidation zone of veins containing iron minerals. It is usually formed by the oxidation of minerals, such a sklerite, pyrite and magnetite. Goethite is a principal constituent of limonite.

Siderite : $FeCO_3$

Crystal System – hexagonal. Cleavage – perfect rhombohedral. *H* - 3½-4. Sp.gr. – 3.96. Lustre – vitreous. Colour – light to dark brown. Habit – rhombohedral crystals. Also massive, granular, fibrous, botryoidal and earthy.

Occurrence. Massive siderite occurs as large sedimentary deposits. It also occurs as replacement deposits formed by the action of ferrous solution on limestones, siderite is also found in hydrothermal metallic veins.

Pyrite : FeS_2

Crystal System – cubic. Fracture – conchoidal. *H* – 6-6½. Sp.gr.- 5.02. Lustre – metallic. Colour – brass yellow. Streak – greenish or brownish black. Habit – common forms are the cube which often show striations. Also massive, granular and stalactitic.

Occurrence. Pyrite is the most common sulfide mineral. It is found in hydrothermal veins, contact metamorphic deposits and as an accessory mineral in sedimentary rocks. It is often associated with chalcopyrite.

Pyrrhotite : FeS

Crystal System – hexagonal. Cleavage – absent. Fracture – uneven *H*-4. Sp.gr. – 4.58-4.65. Lustre – metallic. Colour – bronze yellow tarnishing to brown. Streak – black. Special character – magnetic. Habit – tabular crystals. Also massive or granular.

Occurrence. Pyrrhotite is formed by magnetic segregation in basic rocks. It is also found in high temperature sulfide veins and replacement bodies.

Marcasite : FeS_2

Crystal System – orthorhombic. Cleavage – present. *H* – 6-6½. Sp.gr.- 4.89. Lustre – metallic. Colour – pale bronze-yellow. Streak – greyish black. Habit – tabular crystals common. Also in stalactitic and radiating forms.

Occurrence. It occurs as concretions in sedimentary rocks.

Manganese Minerals

Braunite : $(Mn_3Si)_2O_3$

Crystal System – tetragonal. Cleavage – perfect. Fracture – uneven. *H* – 6-6½. Sp.gr. – 4.75-4.82. Lustre – submetallic. Colour- brownish Mack to steel gray. Streak – brownish black to steel gray. Habit – massive, granular.

Occurrence. Braunite occurs in hydrothermal veins with other manganese oxides. It also occurs with pyrolucite and psilomelane as a secondary mineral.

Pyrolusite: MnO_2

Crystal System – tetragonal. Cleavage – perfect. *H* – 1-2. Sp.gr.- 4.75. Lustre – metallic to dull. Colour and Streak – iron black. Habit – as reniform sooty masses which soil the fingers. Rarely as elongated crystals.

Occurrence. Pyrolusite appears similar to psilomelane. It is a secondary mineral formed by the alteration of manganite or other primary manganese minerals. It is commonly found associated with braunite, goethite and limonite.

Psilomelane: $(Ba.H_2O)_2Mn_5O_{10}$

Crystal System – orthorhombic. Fracture – uneven. *H* – 5.6, Sp.gr.- 4.4-4.7. Lustre – submetallic. Colour – black. Streak – brownish black, shining. Habit – massive, botryoidal and stalactitic.

Occurrence. Psilomelane is a secondary mineral which is formed by the weathering of manganese carbonates or silicates. It occurs usually with pyrolusite.

Manganite: MnO(OH)

Crystal System – monoclinic. Cleavage – perfect. *H* – 4. Sp.gr. – 4.3. Lustre – submetallic to dull. Colour – steel gray to iron black. Streak – dark brown. Habit – prismatic crystals, often grouped in bundles. Also columnar.

Occurrence. Manganite is a low temperature mineral found in hydrothermal veins. It is often associated with pyrolusite, goethite and baryte.

Copper Minerals[26]

Azurite : $Cu_3(CO_3)_2\,(OH)_2$ or $2CuO_3.Cu(OH)_2$

Crystal System – monoclinic. Cleavage – perfect prismatic. *H* - 3½-4. Sp.gr. – 3.77. Lustre – vitreous. Colour – azure-blue. Streak – pale-blue. Habit – crystalline. Also in radiating groups or massive.

Occurrence. It occurs as a secondary mineral. It frequently alters to malachite.

Malachite : $Cu_2CO_3(OH)_2$

Crystal System – monoclinic. Cleavage – perfect but rarely seen. *H*- 3½-4. Sp.gr. – 3.9-4.03. Lustre – vitreous in crystals, silky in fibrous varieties and dull in massive varieties. Colour – bright green. Streak – pale green. Habit – usually in radiating fibres forming botryoidal or stalactitic masses. Also granular or earthy.

Occurrence. Malachite is a secondary copper mineral found in the zone of oxidation of copper veins.

Cuprite : Cu_2O

Crystal System – cubic. Cleavage – poor. Fracture – uneven. *H* - 3½-4. Sp.gr. – 6.14. Lustre – submetallic. Colour – dark red. Streak – brownish red. Habit – often massive, granular or crystalline.

Occurrence. Cuprite is found in the upper oxidized portions of copper veins. It is often associated with native copper, malachite and azurite.

Bornite : Cu_2FeS_4

Crystal System – cubic. Cleavage – absent. Fracture – uneven. *H*- 3. Sp.gr. 5.06 – 5.08. Lustre – metallic. Colour – copper red, tarnishing to purple iridescent. Streak – grayish black. Habit – usually massive.

Occurrence. Bornite is commonly found in hydrothermal ore deposits.

Chalcopyrite : $CuFeS_2$

Crystal System – orthorhombic. Cleavage – poor. Fracture – conchoidal. *H* - 2½-3. Sp.gr. – 5.77. Lustre – metallic. Colour and Streak – dark gray. Habit – short prismatic crystals. Also massive.

Occurrence. Chalcopyrite is formed by the alteration of primary copper sulfides in zones of secondary enrichment.

Covellite : CuS

Crystal System – hexagonal. Cleavage – perfect giving flexible plates. *H* - 1½-2. Sp.gr. – 4.6-4.76. Lustre – metallic, massive varieties dull. Colour – indigo blue. Streak – dark gray, shining. Habit – platy crystals, usually massive.

Occurrence. Covellite is found in the zone of sulfide enrichment. It is associated with other copper minerals, such as chalcocite, chalcopyrite, bomite, etc.

Lead and Zinc Minerals

Galena: PbS

Crystal System - cubic. Cleavage – perfect cubic. Fracture – stepped fracture. *H* - 2½. Sp.gr.- 7.4 -7.6. Lustre – bright metallic. Colour and Streak – lead grey. Habit – often cubic crystals. Also massive.

Occurrence. Galena is commonly found in hydrothermal sulfide veins associated with sphalerite, pyrite, chalcopyrite, etc. It is also associated with silver minerals.

Sphalerite : ZnS

Crystal System – cubic. Cleavage – perfect. Fracture – conchoidal. *H* - 3½ - 4. Sp.gr.- 3.9 – 4.1. Lustre – resinous. Colour – yellow, brown to black. Streak – brown to yellow. Habit – crystals occur as tetrahedral. Also massive or fibrous.

Occurrence. Sphalcrite occurs in hydrothermal veins. It is often associated with galena and other sulfides.

Miscellaneous Ores

Chromite: $FeCr_2O_4$

Crystal System – cubic. Cleavage – absent. Fracture – uneven. *H* - 5½. Sp. gr. – 4.6. Lustre – metallic in submetallic. Colour – black. Streak – dark brown. Habit – commonly massive or granular.

Occurrence. Chromate occurs in igneous rocks such as peridotites and other ultrabasic rocks and in serpentines derived from them. It is also found in alluvial deposits.

Ilmenite : $FeTiO_3$

Cyrstal System – hexagonal. Cleavage – absent. Fracture – conchoidal. *H* - 5½ - 6. Sp.gr. – 4.7. Lustre – metallic to dull. Colour – iron-black. Streak – black. Habit – thick tabular crystals, often massive or as sand grains.

Occurrence. Ilmenite occurs as an accessory mineral in igneous rocks such as gabbros, diorites and anorthosites. It is also found in ore veins and pegmatites. Beach sands may also contain Ilmenite.

Notes and References

1. Previously it was thought that the modern man first appears in the fossil record during the later part of the Upper Pleistocene around 35,000 b.c. from Niah Cave in Sarawak, East Malaysia, but now the earliest date of *Homo sapiens sapiens* in South Africa dates back to *circa* 1.5 lac years ago and in India *circa* 75,000 years ago. Albeit the dates of Upper Palaeolithic culture of Karnataka and central india goes to 25,500 --10,500 years ago.
2. The earliest epoch of the Quarternary period, starting some time around 2 million years ago and ending at *circa* 10,000 years ago. The Pleistocene was succeeded by the Holocene or recent epoch.
3. The lower Palaeolithic had the predominance of core tools of pebble tools, handaxe and chopper chopping which formed by Clactonian and Levallois technique with the earliest forms of man (*Australopithecus and Homo erectus*).
4. The quartzite is common in the Archaean and Dharwar and the older Purana system is a granulase, recrystallised metamorphic rock, composed essentially of quartz with or without sericite, rutile or other accessory constituents. It may be derived from original siliceous sediments or from quartz-reefs and vein-quartz. In absence of stratification planes, ripple marks and other sedimentary characters, it is difficult to distinguish sedimentary from igneous quartzite. It is most common rocks of the Cuddaph system, from the occurrence of the most typical and first-studied in the district of Cuddapah in the middle of Tamilnadu. Quartzites are too hard to work and have a fracture and grain unsuitable for dressing into blocks.
5. The Middle Paleolithic was the era of Neanderthal man and the predominance of flake-tool industries prepared by Mousterian technique, was technically more advanced than anything which had gone before in the form of Levallois technique.
6. The Upper Palaeolithic (starting perhaps as early as 38,000 B.C., with *Home sapiens*, blade-and-burin industries.
7. The Mesolithic was a period of transformation between Palaeolithic and the Neolithic, distinguished by an abundance of microliths.
8. Chert is a very finely crystalline form of the mineral silica, found as nodules in limestones. Many cherts are fine-grained that they behave like a glass—they fracture almost as sharply as obsidian and may be chipped or flaked to make artefacts.

9. Chalcedony is a very finely crystalline form of the material silica, rather similar to chert. They are many varieties, several of which are prized as semi-precious stones, for example agate, onyx, carnelian and jasper. They may be used in making beads and other jewellery.
10. The mass of a material per unit volume is called density. Where a large number of small items have a parameter which itself varies over a large range, it is sometimes useful to express the number of items per unit interval of the parameter as a density of items as a function of the parameter. In the case of density the common feature is that the interval over which the items are counted must be much larger than the items themselves, but much smaller than the range of parameter.
11. Periodic table of the elements listed in order of increasing atomic number was the great discovery in chemistry when Dmitri Ivanovich Mendeleyev (1834-1907) tried to organise the elements, he knew into a coherent pattern that made chemical scale. The horizontal rows of the table are called periods and the vertical columns are called groups.
12. Mohś scale is a qualitative scale of hardness invented by Friedrich Mohs (1773-1839) in which a substance with a higher number on the scale can scratch all substances with a smaller number. Talc is given the value 1 on the scale, and diamond 10.
13. The emission of light by a substance which is itself exposed to electromagnetic radiation, especially light is called as fluorescence. The incident light excites electrons to higher electronic energy levels, and the subsequent return to the ground state causes fluorescence. The whole phenomenon of fluorescence colouring in minerals are of within about a microsecond of stimulus ceasing.
14. It is a form of luminescence characterised experimentally by a delay between the removal of the source of illumination and the cessation of emission from the phosphorescent object.
15. Felspar (Feldspar) is a group of minerals that consist of silicates of aluminium, combined with potassium, sodium, calcium and barium. It forms one of the most common minerals in igneous rocks. When it is dominated by potassium then it is grouped under alkali felspar, while with sodium or calcium it is plagioclase felspar. The crystals are white, pink or grey.
16. Silicate is a group of minerals based around the highly stable SiO_4, constituting some 95 percent of the earth's crust and including clay minerals, Felspar, Quartz.

It occurs in cryptocrystaline form (as chalcedony) in amorphous and hydrated forms (as opal), and combined form in silicates.

17. The minerals crystallize in both orthorhombic and monoclinic systems and all species have similar cleavage. The principal orthorhombic members of the group are enstatite and hypersthene; the monoclinic members are diopside, augite and jadeite.
18. It is a large group of rockforming metasilicate minerals have a structure of silicate tetrahedra linked to form double endless chains, in contrast to single chains of the pyroxenes, to which they are closely related. They are present in many igneous and metamorphic rocks. The amphiboles show a wide range of compositional variation but confirm to the general formula: $X_{2-3}Y_5Z_8O_{22}(OH)_2$, where X = Ca,Na,K, Mg or Fe^{+2}; Y = Mg, Fe^{+2}, Fe^{+3}, Al, Ti or Mn; and Z = Si or Al. The hydroxyl ions may be replaced by F, Cl or O.
19. A group of transparent silicate minerals common in Acid Igneous Rocks, in Metamorphic Rocks, and in derived sediments, which having perfect cleavage, can easily be split into very thin form.
20. It is a group of rock-forming silicate minerals, usually dark green, consisting of silicates of Magnesium and Iron, occurring in Basic and Ultra Basic Rocks and probably a major component of the SIMA.
21. A generic name for a group of mineral silicates that are isometric in crystallization; have a structure consisting of independent SiO_4, tetrahedra and AlO_6 octahedra, which are linked by cations in 8-coordination.
22. It is a hydrous ferrous iron magnesium aluminosilicate; may also have ferric iron (Fe^{+3}), chromium or manganese. It occurs in monoclinic or pseudohexagonal form silicate sheet structure as micas. It shows basal cleavage in crystallized forms.
23. A group of common rock forming minerals containing many polytypes almost always occurring as fibrous yellowish-white or green aggregates. It decomposes at high temperatures to forsterite, talc and water.
24. It is a finely form of crystalline, hydrous magnesium or aluminium silicates forming the major constituents of clay; crystal structure of the two layer or three layer type. Kandite group have two layer structure and composition $Al_2O . 2SiO_2 . 2H_2O$ with ferrous iron and magnesium substituting for aluminium. Three-layer types have two layers of silica tetrahedral interleaved with alumina di and the octahedral.

Many clay minerals are of mixed-layer types of randomly or regularly interstratified growth of two or more clay minerals.

25. Iron is a widely occurring and the second most widespread mineral, aluminium being the first, estimated to constitute chemically 5 per cent of the earth's crust by weight. It does not occur native in nature due to its high oxidation number by which it is chemically active and forming compounds.

26. It is mainly obtained from a variety of ores occur especially in the veins or scattered through metamorphic rocks. Copper usually does not contain any isomorphic impurities and its crystalline structure is analogous to the structure of gold and silver. Its origin is said to be exogenetic when copper is mainly found in the oxidation zone of copper deposits in association with chalcocite, cuprite, bromite, limonite and calcite. Its hydrothermal origin is rare, besides the latter is always of low temperature.

CHAPTER 3

COPPER METALLURGY

Copper is a salmon-pink and highly ductile metal occurs in native form contain Ag, Bl, Pb etc. and some of their minerals used as a natural pigments. It is also one of the few base metals like iron which is found in its natural state in quantities large enough to be worked in this way, and therefore, like gold, it is one of the first metals ever used by man. Since it is a base metal means it can readily form oxides or sulphides in the nature therefore, its principal ores are maladite, $CuCO_3.Cu(OH)_2$, azurite, $2CuCO_3.(CuOH)_2$; cuprite, Cu_2O; melaconite or tenite, CuO; copper pyrites, $CuFeS_2$ and erubescite, Cu_2FeS_3. The most important family of alloys based on copper are the bronze and brass. Here bronzes is a general term originally reserved for copper based copper-tin alloys, in which low tin contents of the order of 3-4% provide wrought materials suitable for war and agricultural implements in the yore, while tin contents of 10% or so give good casting properties e.g. Gunmetal, and tin contents of 15 to 18% provide a hard, sonorous, castable and corrosion-resistance alloy used for bells. The brasses are essentially a copper-zinc alloys containing upto 50% Zn. Thus in modern connotation, bronze is a copper-rich alloy other than a brass e.g. in gunmetal 85% Cu, 5% Sn, 5% Zn and 5% Pb is present.

The Neolithic and early Chalcolithic pyrotechnologies included the firing of clays to make pots and the heating of stone materials to enhance colour, workability and hardness. Possibly fire setting was being used to extract native copper lumps and granules that could be further processed by hammering and annealing. Thus the discovery of metal smelting and melting could be result of intentional experimentation. An example is the site of Mehargarh which was excavated by the French Archaeological Mission to Pakistan under the direction of Jean-Francois Jarrige from 1974 to 1986 and again in 1996-97, where during the 5th to 4th millennium BCE, changes were occurring simultaneously in metal production, so stone working, production of fired and steatite beads and shell working.

Copper is relatively a soft metal, which can be cold worked and melts at a low enough temperature to be cast using the technology available in antiquity. During cold working plastic deformation in a metal started at a temperature low enough to result in permanent strain-hardening. In practice, the metal may take shape at a temperature well below the recrystallisation temperature (since as the degree of plastic deformation is increased, the temperature at which subsequent recrystallisation can occur is decreased). The metal occurs only 0.5% of the earth crust[1] rarely, as nuggets of native copper; it is in this form that it was first worked by man. More frequently, copper is obtained from ores. These are quite widely found in the form of oxide ores (cuprite), sulphide ores (chalcopyrite, chalcocite and bronite) and carbonates (malachite and azurite). In nature, copper deposits are present in igneous[2] , sedimentary[3] and also from metamorphic[4] rocks. The ores generally occur as veins[5] , disseminations and as bedded[6] deposits. The latter two form the important sources of copper in the world. In India, the copper deposits are mainly in Jharkhand, Rajasthan, Andhra Pradesh, Karnataka, Tamilnadu and in the Garhwal-Kumayun belt of Uttaranchal, in Sikkim, Kumaon and Nepal. In the Chhotanagpur plateau of Bihar and adjacent West Bengal and Orissa copper ores have a wide dissemination but the most significant occurrence lies in the Singhbhum-Hazaribagh belt of Jharkhand. Old workings abound, but the following localities deserve special attention—Baragunda (24^0 4'N; 86^0 7'E) in Hazaribagh and Rakha (22^0 38'N; 86^0 27'E)—Mushabani (22^0 31'N; 86^0 28'E) localities in Singhbhum. Baragunda is the site of 48 ancient copper mines which extended all along the outcrop and average 2-3 km in width along the main Singhbhum-Hazaribagh line. The major smelting area seems to have been Giridih, about 3 km away, where large mounds of slag covering several acres of ground may be observed. The richness of the old workings in and around the Rakha mines will be apparent from the fact that between the Rakha mines and Roam, an area of several acres was found to contain slag to a depth of 3.5 meter. F.R. Allchin based on his study of D.H. Gordon collection, report the presence of 2nd millennium BCE Chalcolithic pottery of Eastern Indian feature is reported near Rakha mines. In this context, it must also be noted that old copper workings occur even outside the Singhbhum-Hazaribagh belt and some of them are found well outside the known copper-bearing. For instance, in West Bengal, in Bankura District Dilip K Chakrabarti reported remains of at least two old copper mines—one at Chhedapathar where one notes old shafts for about 1 km and the other at Tamkum where the remains of an open shaft into the side of a hill are clearly visble. The Singhbhum copper belt in Jharkhand is a part of Singhbhum shear zone of about 140 kms. in length. This belt, commencing at Duarapuram with an E-W trend, passes through Khagswan and Seraikella from where it takes a turn to the Rakha mines and Musabani, ending in region south east of Bahragora. The host rocks, which are intensely deformed, belong to the Precambrian Iron-ore Series comprising biotite-chlorite schists, mica-schists, phyllites, quartzites, epidiorites and hornblende schists. Migmatization and ore-localization are the main features within this shear zone and they are thought to be genetically related. Two phases of mineralization have been reported. The earlier phase contribute to the formation of oxide minerals apatite $[Ca_3 (PO_4 .CO_3)_3 (F, U, OH)]$-magnetite ($Fe_3O_4$)veins and uranium bearing veins while the latter to the formation of copper sulphide minerals. The presence of apatite confirm its igneous origin.

Recent studies indicated that the uranium mineralization followed the apatite-magnetite formation but was prior to sulphide mineralization. Therefore, in this area copper sulphide ores is present above to apatite-magnetite deposit which can be seen at Mursabani and Rakha mines. In this area the copper ore is chalcopyrite which occurs in mesothermal deposits associated with other sulphides like pyrrhotite, pyrite, pentlandite, nickel pyrrhotite.

In Rajasthan the copper mines areas are mostly in the east of Aravalli (Mewar region) and not in west of it (Marwar region), extending from Bharatpur, Alwar and Khetri regions in the northeast to south of Udaipur in the south-west reaching as far as Amba Mata on the border of Rajasthan and Gujarat. The old working localities which have been reported from Bharatpur are Nithahar (26° 58'N; 77° 06'E), Baswar (27° 02'N; 70° 07'E), Hathori (27° N; 77° 10'E), Khankhera (26° 58'N; 77° 06'E) and Khareri (26° 11'N; 77° 56'E). A more than 3 meter deep it has been reported at Khankhera and old slag heaps are said to occur at Khareri. In the Alwar area the most impressive occurrence of old mines may be seen at Khoh Dariba (27° 10'N; 76° 27'E) in the form of vertical shafts which definitely go up to 60 meter in depth. At one point the circular mouths of these shafts extend intermittently for about 1 km. Another important locality in this zone is Pratapgarh (27° 15'N; 76° 13'E). The strike length is about 10-11 km and falls into two blocks—Madhupura block (northern) and Nagel block (southern). There are 200-300 closed-space vertical shafts, some going upto 90 meter in depth in this locality.

The Khetri copper belt in Rajasthan extends about 80 kms from Singhana (Junjhunu district) to Raghunathgarh (Sikar district). Mineralization has taken place in phyllites[7], schists[8] and slates[9] of the Delhi-Aravali system.[10] The ore localization in this is thought to be primarily controlled by structural features (shear zones and fractures). The ore bodies occur an enechelon ore-shoots, stringers and also as disseminations. In this belt, the Kolithan and Khetri prospects are in chlorite schists while the host rocks around Dhanaota are phyllites. In Khetri belt, pyrite, pyrrhotite and chalcopyrite are the main sulphide ores along with cobalt sulphides. The common feature in Khetri of Rajasthan and Singhbhum of Jharkhand is that the copper ores of northern region occurs as veins or as dissemination in the Dharwar schists and phyllites. The Khetri belt is the most well-known copper bearing belt in Rajasthan and well known for its old workings. In Khetri (28° N; 75° 51'E) old workings are found near the top of a ridge, some 500 feet above the present plain. There is an underground gallery known to be more than 5 kms long, following the strike of the formations. Some deep shafts connect with underground gallery. At Babai (27° 53'N; 75° 49'E) in the same area the workings are quite irregular, commencing as narrow inclines, widening out into large chambers, contracting to narrow burrows, as the ore led the miners, narrow burrows, and branches as the as the ore led the miners, but their general direction is with the dip of the slates at an angle of 30^0 to 60^0. Perhaps the most important remains can be seen at Madhan mine at Khetri. The entrance to the upper part of the mine is on the top of a hill where the subsidence of 40 meter in depth over an area of 80 meter by 120 meter has been exposed which can be reached by a 4 meter by 2 meter opening. The scale of mining operations in this area is also exemplified by a huge slag deposits looking like hillocks at Singhana (28° 06'N; 75° 54'E) and also by the innumerable smaller slag heaps on the bank of Kharkera river. Singhana was the most important smelting center of the area. A little away from Khetri is the old mining area of Baleshwar near Neem-Ka-Thana. In the bills overlooking the valley there must be more than 100 old workings both in the form of vertical shafts and open trenches dug into the hillside and the whole area is dominated by rubbish material deposit of slag. But so far the actual copper working in the Khetri belt have not been dated beyond 1000 B.C.E.

Table of Radiocarbon dates of Ancient Copper Workings in Rajasthan and Gujarat

S. No.	Site	Sample No.	Locus	Radiocarbon date in B.P.
1.	Khetri Rajasthan	PRL-929	Depth 40-50 m.	250 ± 80
2.	Alwar Rajasthan	PRL-927	Old mine shaft	Modern
3.	Jhunjhunu Rajasthan	PRL-684	Ancient mine	Modern
4.	Dariba Mines Dt. Udaipur, Rajasthan	TF-1117 Layer 436	Depth 64 m,	2310 ± 105
5.	Kumbaria Dt. Banaskantha,Gujarat	TF-1221 Jain Temples Kumbaria	Dump behind	535 ± 90
6.	– do -	TF-1222	- do -	905 ± 85
7.	Ambamata Dt. Banaskantha Gujarat	PRL-53	1L m distance in No. 3 West Drive in the underground Mine	2110 ± 200
8.	– do -	PRL-66	Bone hole No. 100 of Ambata deposit	850 ± 100
9.	Dariba mines Dt. Udaipur, Rajasthan	PRL-208	East Lode old working depth 4 to 5 m.	2200 ± 100
10.	– do -	PRL-210	Main Lode old working depth 100 m.	1840 ± 130
11.	– do -	PRL-209	East Lode old working depth 263 m.	3120 ± 160

In Andhra Pradesh the copper ores in the form of chalcopyrite occurs as veins, stringers and disseminations at Agnigundala (Guntur district), Bondalamottu, Nallakonda, Dhukonda and Garimanipenta (Nellore district) and Gani-Kalana areas (Kurnool district). The copper deposits are confined to calcareous quartzites with appreciable amounts of cobalt, nickel and silver. Geochronologically, the copper ores of Rajasthan are probably of Delhi age, while those of Andhra Pradesh are of Cuddapah to post Cuddapah age. Here Delhi is of Archoeozoic while Cuddapah belonging to the Proterozoic era of the Precambrian time consist of sedimentary rocks which have been subjected to varying grades of metamorphism, but the Delhi system has undergone more severe folding in contrast to Cuddapah system. Chronologically Delhi system is slightly earlier than Cuddapah system. Economically the Cuddapah system of Andhra Pradesh is dominated by iron and manganese ores, while Delhi system of Rajasthan is by copper and some cobalt and nickel ores. Due to this very fact most of the chalcolithic culture were flourished in the catchment area of Delhi system, while iron age culture in the catchment area of Cuddapah system in India. In Karnataka, the Ingladhal deposits around Chittradurga district occur in ferruginous chert bands intercalated with the greenstones. Here the mineralization appears to be structurally localized. Deposits also occur around Kalyadi (Hassan district) in quartzite bands in schists and gneisses. The ore occurs as lenses, stringers and disseminations of chalcopyrite within these quartzite. In Tamil Nadu copper deposits are reported from Mamandur in South Arcot district. The ore is localized in the shear zone within gneines and charnockites. Native copper occurs at some places in South India. In Kashmir isolated masses of pure native copper have been found in the bed of the Zaskar river[11], but their source is unknown. They occur there as water-worn nodules weighing up to 10 kgs.

Archaeologically India has had a long history of copper metallurgy by the time of Early Harappan Culture (5500 – 2600 B.C.E.) whose mature phase possibly began around 2600 B.C.E. and the tradition continued right through the subsequent phases of the Indian protohistoric and historic sequence. Although iron appeared sometime in the first half of the second millennium B.C.E., but the remains of copper technology continued and flourished during pre-industrial time too. How the neolithic pattern on early food producing era[12] (7000 – 5500 B.C.E.) transformed itself into the copper-bronze age and ultimately led to the formation of the mature Harappan civilization or transformation of regionalization era (5500-2600 B.C.E.) into integration era (2600-1900 B.C.E.) was an important issue that attracted the attention of American, British, French, German, Italian, Indian, and Pakistani archaeologists. The early work done by Sir Aurel Stein[13], H. Hargreaves[14], Beatrice de Cardi[15], Water A. Fairservis, Jr.[16], George F. Dales[17] showed the importance of Baluchistan because it was an intermediate region between the Mesopotamia and Iranian complex to the West and the great region of the Indus to the east, it was only natural to build a working hypothesis for discovering a link between the two areas in the upland of Baluchistan. The work of J.M. Casal[18] at Mundigak in southern Afghanistan revealed a sequence that was relevant to the study of the Baluchistan early Harappan materials. But the excavations at Mehargarh in the Kachhi plain in south Baluchistan by Jarrige and Lechevallier[19] have revolutionalized our whole concept of the Harappan Bronze Age culture in Pakistan and India. All past attempts to trace and derive the Harappan Bronze Age culture from the west have been surpassed. Here we have beginnings of the Bronze Age emerging at almost the same time as in the west. Moreover, they have their own particular characteristics, suggesting that neolithic in the Indus region move on to the bronze Age by a natural process of evolution without any outside interference of the west.

There are the three important sites of Mehargarh, Nausharo, and Pirak which together make up a continuous history of the eighth millennium B.C.E. down to the early historic period, thus enabling us to observe an uninterrupted succession of cultures in the Kachhi plain, a transitional area between two different worlds of the arid inland plateaus of Baluchistan, Afghanistan and Iran on one side and the Indus plain on the other. Another site of Sibri, not far from Mehargarh, has now been explored. It dates from the end of the third to the beginning of the second millennium B.C.E. by Jarrige.[20] While Pirak and Sibri have supplied the material for the post-Harappan phase, Nausharo gives the mature Harappan sequence, and Mehargarh provides the antecedents of the Harappan Civilization in the

Indus plain. It may be noted that the Nausharo mound yielded on the surface 80% typical, mature Harappan objects together with a few fragments of human figurines, which correspond in their fabrics and styles to specimens from the uppermost levels of period VII at Mehargarh.[21] Considering the fact that Mehargarh no mature Harappan finds have been noted, except for perforated pots on the surface, Mehargarh was probably deserted before mature Harappan colonization.

The copper ore deposit of Baluchistan play a vital role in the transformation of neolithic culture to chalcolithic culture in yore. The Las Bela, Jhalawan and Chagai districts of Baluchistan have high amount of copper deposit. The *Mazhar-i Shahjahani* of Yusuf Mirak which was compiled in 1634 A.D. mention copper deposits of Las Bela region as cited by Irafan Habib[22] in 1982. The District Gazetteer of Lal Bela published in 1907 mentions, in this regard, the account of an Englishman called S.V.W. Hart[23], a Captain of the 2nd Bombay Grenadires, who visited the area as early as 1840:

> "In the hills between Liari and Bela copper is found in large quantities, but from the dread of exciting the cupidity of the neighbouring chiefs, it is not allowed to be worked.... The whole country is indeed rich in mineral productions and well worth the attention of an experienced geologist."

The copper deposits were apparently, in the vicinity of the Kunnaraj river and Bela and in 1840, the refined copper of the region was called as *Kattoo* which was extracted by simply melting of copper ore with wood in a mud furnace and the ore ran off like a stream of gold.[24] The Wad hills of Jhalawan district have large quantities of malachite contains 57.3% copper and in the Saruna and Khidrani area in the same district have rich chalcopyrite containing 36.4% copper associated with little malachite is well known by the Britiohers.[25] A detailed account of the copper resources of the Chagai district, provided by E. Vredenberg, is available in the Bolan and Chagai district gazetteer of Buller.[26] At Ras Koh and Las Koh chalcopyrite occurs, while at Robat Koh, Malik-i-Siah Koh and Saindak Koh in the same district carbonate of copper occur in some veins. Specimens from the Saindak Koh now called as 'Sandak Copper mines'[27] in the Chagai district have been dated to around 3500-3000 B.C.E. Also, at Robat, near Lar Koh, Vredenberg noticed significant heaps of slag which were so numerous that a small fort had been built entirely by them. Therefore, it is clear that some copper may also have been procured by the early Harappan peoples of north and central Baluchistan from the region to the South, in the light of the impressive copper ore deposits there, especially in Las Bela area. On its western side, the possibility of the Chagai district being a supply area for copper to the early Harappan culture of the Quetta region. The Nal and Kulli types of early Harappan settlement are concentrated mainly in the Las Bela region of Baluchistan due to its high potentiality of copper ores.

Intensive archaeological exploration in the Baluchistan has advanced our knowledge of settlement patterns and site density in some of the key regions where copper ore has been located. Data can now be presented on site locations, numbers and sizes from the beginning of early food producing Neolithic era (7000 to 5500 B.C.E.) in the form of Mehargarh phase to early Harappan chalcolithic era (5500 to 2600 B.C.E.) in the form of Hakra, Kechi Beg and Balakot phases by Rafique M. Mughal as follows:

Phase	Site counts Measured	Settled area in hec.	Average size in hec.
Early Harappan c. 3200-2600 B.C.E.	477	2146	4.5
Kechi Beg/Hakra c. 5500-3200 B.C.E.	256	1331	5.2
Mehargarh c. 7000-5500 B.C.E.	20	52	2.6

The above data seem to show a strong and continuous growth of settlements and population in the area of Baluchistan from Neolithic to Chalcolithic i.e. from Mehargarh phase to early Harappan phase only due to the changes from lithic to copper technology by the locally available copper ores in the form of carbonate, oxide and sulphide. Thus the Mehargarh chronology (from sixth millennium B.C.E. to middle of the third millennium B.C.E.) provides a time scale of different cultural patterns, which may be seen as the antecedents of the Harappan civilization. It also allows us to see the evolution of this great urban civilization in a different light of the local material culture which provides a substantial corrective to earlier theoretical assumptions.

While Baluchistan was not bristling with mature Harappan sites, the work of George Dales[28] at the sites of Sutkagen-Dor and Sotka Koh on the Makran coast and Balakot in the Karachi Bay[29] and the recent excavations at the site of Nausharo six kms. from Mehargarh, by the French Archaeological Mission from 1974-86 under the leadership of F.J. Jarrige clearly indicates that Baluchistan was not a peripheral region in any sense. High potential centers such as Nausharo, Dabarkot and Balkot were flourishing Harappan remains have been also noted at Periano Ghundai, Kaunri, Sinjawi, Spina Ghundai and Quetta Miri by Mughal.[30]

In fact, as far as Sind is concerned, it was very much a part of the interregional network of south and central Baluchistan since from the fourth millennium B.C.E. with Nal and its related sites sharing with the Amri-related horizon a striking similarity in its ceramics. The Gomal valley of Baluchistan is geomorphologically functioning as a funnel between north Baluchistan and the Indus plains was manifested as early as the third millennium B.C.E. with the presence of the Kot Dijian complex, intermediate between the Baluchi cultural developments and the mature Harappan civilization at both Gumla and Rehman Dheri. Some raw materials that were exploited by Gumla and Rehman Dheri like copper have come from Baluchistan. The whole Indus plain from Harappa in the north to Mohenjodaro in the Sind were located in a minerally poor area especially in context of copper ores. Although Ball[31] and Wynne[32] of the Geological Survey of India had noted copper in the salt range of Jhelum District which were called copper shales owing to the presence of small nodules of copper sulphide which were not of great economic value. The copper artifact from Mohenjodaro, Chanhudaro, Allhadino in Sind in the Indus plain testify to the requirements of the region for ore source which could not be obtained locally, but which were available in Baluchistan in the form of sulphide ore. However, Baluchistan was not the only region with which Sind was interacting for copper ores and it is mostly likely that of Rajasthan and Gujarat were also involved in various kinds of trade relationships with it. Considering that the chemical and metalographic analysis of bronze artifacts of Mohenjodaro clearly shows that heterogeneous sources of copper ore were being utilized, and also keeping in mind the existence of rich copper deposits in Rajasthan whose cultural complex contemporaneous with the early and mature Harappan culture i.e. the Ganeshwar Complex[33], shows evidence of contacts with the latter, it is most likely that copper ore was being imported from Rajasthan. The recent discovery of an early Harappan Copper Age culture in the region around Ganeshwar in Sikar district of Rajasthan is decidedly a landmark in the protohistoric archaeology of India. This Ganeshwar-Jodhpura complex spreads towards the Harappan sites in north-western Rajasthan through the Kantli River. This fact also suggest that the Kasaunti-Dohan-Sabi and Sota rivers helped these people in their movement towards Haryana in Harappan and late Harappan phase. It is very likely that the copper mines around Khetri in Jhunjhnu district, Neem-Ka-Thana in Sikar district along with Alwar-Jaipur copper belt was the source of copper for Kalibangan too which is located 200 kms north of it.

To trace the characteristics of the metals and sources of ores used for manufacture of copper-bronze implements in Harappan culture, scholars like Sanahullah, Agrawal, Hegde, Bharadwaj have tried to correlate the impurity patterns of ores and artifacts from different Harappan sites. Sanahulla[34] gave importance to the presence of arsenic in Harappan artifacts and tried to distinguish them from contemporary Mesopotamian objects.

Agrawal[35] used the methods of statistical approach suggested by Friedman and Forbes to arrive at the relative probabilities of occurrence of impurities in the case of native oxide and sulphide ores of copper and concluded that there is a high probability of extensive use of the oxide ores and the native minerals as compared to sulphide ores. According to him out of the 177 artifacts analysed earlier from Mohenjodaro and Harappa, only 30% were alloyed. Tin alloying ranged from 1-12%; arsenic alloying 1-7%; nickel alloying 1-9% and lead 1-32%. Tin bronzes were more common than any other alloys. The Harappans knew the technique of striking 'rising', 'running-on', cold work, annealing, riviting, lapping closed casting, Cire perdue etc. The sulphide ore smelting of copper was perhaps known right from the early Harappan phase as a result of diffusion. Spectroscopic analysis of various Indian copper ores and two Harappan artifacts have been reported which suggest that only the Khetri ore has close correspondence with impurity pattern of Harappan artifacts. Unfortunately, although the copper reserves and old workings of Rajasthan were known as exploited from an ancient period there is a great paucity of scientific dating. Regrettably the available radiocarbon dates for Khetri are only *Circa* 1000 B.C.E. to medieval period. But the presence of Harappan pottery in the Sikar and Jhunjhunu districts of Rajasthan where Khetri copper belt is present and the occurrence of copper celts at Sabania in Bikaner with Harappan script confirms that the Harappan demand for copper came from Khetri copper belt of Rajasthan. Here it is pertinent to point out that Bikaner is just west to Jhunjhunu on 28^0 east-west longitude in the same ecological zone. Recently six copoer-ore mining and smelting sites were investigated for lead isotope dating in the Aravallis although the results are not available as yet. Hegde also reported the dating of timber supports from a gallery set at a depth of 120 meter at the Rajpura-Dariba mines in the Udaipur district, the radio-carbon date is 3120 B.P. i.e. the last quarter of the second millennium B.C.E. In Rajasthan, at Kalibangam, the large number of copper/bronze artefacts which have been found were manufactured in all probability from copper produced from Rajasthan.[36] The existence of copper ore in Gujarat have been reported from

Panchmahals and Sabarkantha district, along with a copper belt of Devi-Ambamata of 8 to 10 km and Kui-Chitraseni belt of 45 kms. extended from Banaskantha district of Gujarat to Sirohi district of Rajasthan.[37] Radiocarbon dates and literary references testify to the existence of an ancient copper mining industry in the Ambamata area from about the first century B.C.E. to the 11th-15th centuries A.D. when it was in a flourishing state, but Harappans of Gujarat did not exploit the copper ores of Gujarat, but their source of copper must be different from the copper ore source of Mohenjodaro and Harappa because at both the sites copper contained nickel and arsenic, while Lothal copper is arsenic free. This suggest on the basis of arsenic as impurity, the copper at Mohenjodaro and Harappa may also have come from Khetri area of Rajasthan (sulphide ores of copper) while Lothal copper came from south Rajasthan (oxide ores of copper which is arsenic free). At Lothal an ingot of copper from a workshop of a metal worker was reported by Rao and by him this bun-shaped copper ingot came to Gujarat during Harappan period from the copper mines of Oman and the Persian highlands. Thus at Lothal (22° 31'N; 72° 15'E) by his opinion of the copper being of external origin but alternatively it can be argued that this bun shaped copper ingot at Lothal may only represent a stage in the preparation of a metal artifact by local metal worker. It is also important to note here that Oman and the Persian highlands never be a good producer of copper ores (Britanica Atlas, Copper minerals, London, 1984, 305). In an important ethnoarchaeological study of indigenous metalcraftsman in India, Meera Mukherjee refers to the local tradition of preparing metal objects from ingots at Kota Padu in Andhra Pradesh. Therefore, it is clear that the whole Harappan area is not likely to have obtained copper from a single source but they have polycentric source of Baluchistan[38] in Pakistan and Rajasthan in India.

The Cholistan region of Pakistan and the adjacent region of Rajasthan in India which lie basically along the dry beds of Ghaggar of India and Hakra of Pakistan have seen some of the most important protohistoric research in the context of evolution of early Harappan culture from its antecedent Hakra culture to the evolution of mature Harappan culture in the recent past. Mughal's work in Pakistan[39] has led to substantial modification in the Harappan settlement patterns. On the Indian side, one of the largest Harappan settlements of the country i.e. Kalibangan (22° 05'N; 74° 05'E) is located here along with a number of other sites in Rajasthan and Haryana.[40] Mughal notice[41] that there is a profusion of industrial sites (47.7 per cent of the total sites) in the mature Harappan phase with a clear separation from the habitation areas in Hakra region of Cholistan definitely used for copper smelting. Also, Siddhanawala Ther in the same area contained numerous kilns, although the use to which they were put has not been started by him. In fact this region of Cholistan played a very significant role in providing copper, both in an unfinished and finished form to other regions of the Harappan area. This specialized industrial activities were being carried out at the late Harappan cemetery-H related complex here is evident from the statistics supplied by Mughal i.e. 18 per cent of the sites were 'exclusively industrial sites' and 28% were 'settlement sites with kilns or specialized activity'.

In Cholistan area fifty sites with cemetery-H related materials are reported by Mughal. This number is far less than that of the mature Harappan period (174 sites) but they are apparently concentrated in the very same area where the mature Harappan sites were located (and where very few early Harappan sites are found whose total number is 40 only). Since the ceramic material of this complex is found from the upper Indus plains to the adjoining region of Punjab and Haryana, the Cholistan tract was obviously very much the region through which copper metallurgy communication was taking place. Thus the significance of the Cholistan tract and the major copper metal technological movement line which followed the present day dry bed of the Hakra river in articulating the people movement between Panjab and Rajasthan in India with regions in north and west in Pakistan is also confirmed. There was also a route which passed from the Sukkar/Rohri hills to Harappa on the one hand and the Ghaggar/Hakra/Nara stretch on the other for copper ores, because Indus plain have no copper ore. There was also a copper ore migration route from Ganeshwar to Sothi-Bhadra via Neem-ka-Thana and Khetri in Rajasthan in which the Kantti river was mediating their migration between Rajasthan and the regions to its north and west as well is likely. This river flowing from Sikra-Jhunjhunu region reached the Ghaggar-Hakra depression at a point between Nohar and Bhadra where it joined the Drishdvati, and from where Kalibangan and other sites to the West were reached.

The Italian Archaeological Mission[42] in association with Peshwar University of Pakistan has attempted to work out the cultural sequence of North-West Frontier area of Pakistan from Neolithic to early historic period exploring and excavating a large number of sites other than Harappan material. Professor A.H. Dani[43] found large protohistoric grave complexes in Timargarh, Balambat and the Thana area as a result of which this cultural complex has somewhat inexactly been described as the Gandhara Grave Culture. The precipitate of copper smelting and the prince of Chisel at Pit No. H2 at Loebanar suggest the possibility of local people in the yore came from Chitral and Waziristan.[44] The Kashmir region of India have a dense distribution of Neolithic settlements in the Baramulla, Srinagar and Anantnag regions from *circa* 2300-1500 B.C.E. on the terrace surface of the Jhelum and its tributaries, known as Karewas. The commonality of a number of ceramic traits between the Kashmir of India and Swat Neolithic cultures of Pakistan is evocative of the interconnection between them. Its links with the Indus region are not very clear although we have drawn attention to the Kot Dijian 'horned deity' at Burzahom (34° 10'N; 73° 54'E) and a singe spiral-headed copper implement at Gufkral. But one thing is clear that Kashmir have a high potential copper ore. In Anantnag district on the slopes of the Mammal and Dunpathri, malachite stains and old working were noted. In Baramulla district pyrite and arsenopyrite have been noted at the Lashteal hill spur, while in Dola district seventeen old workings were also noticed at Duligam area.[45]

During regionalization era of Harappan culture (Early Harappan phase), alloys may have first been used to produce different colour tones in metal, but metalsmiths soon came to appreciate that alloys also increased the hardness and working life of a metal too. On Moh's scale bronzes are slightly less hardened than the iron. The copper-tin equilibrium diagram confirm the pronounced separation between liquidus and solidus results in heavy coring of cast alloys, with the result that in practice if tin content exceeds 7 to 8% then its hardness increases to its double. Although the solubility of tin in copper rapidly decreases below 350° C, transformations are very sluggish below 500° C and the Cu_3Sn stage only appears after long annealing treatment in which Harappans were fully expertise. Smelting of copper ore took place in the distant highland mining region of Baluchistan and the Aravalli region of Rajasthan. Traders or nomads would have brought the distinctive bun-shaped ingots of smelted copper, along with tin, lead and other metals of the skilled metal workers in the urban workshops. Given the fumes, smoke and fire associated with metal processing, many of the metal workshops were probably located along the edges of the settlement or in separate craft villages such as Chanhudaro, where copperworking and lapidary arts were practiced alongside ceramic production, shellworking and possibly even sealmaking. The copper alloyed with arsenic or tin was produced specific objects. Arsenical copper makes extremely hard-edge tools, such as saws or chisels, but the edges were slightly more brittle than those of tin bronzes. The hardened bronze saws of Harappa and Mohenjodaro were as effective as the traditional modern iron-steel saws still used by shellworkers of Gujarat and Bengal. Relatively few copper-bronze objects of Harappan culture have been subjected to compositional analysis, but these initial studies suggest that the Harappans metalsmiths did not maintain a uniform standard for alloying even in their mature phase of urbanization. However, it is also possible that part of the variation in alloy percentages may result from indiscriminate recycling, where objects with different percentages of tin or arsenic alloy were combined to make new object.

The chalcolithic complex of North-East Rajasthan is identified by ochre-coloured pottery which was subsequently christened the Ganeshwar-Jodhpura complex in Sikar district forming the core region due to its heavy copper ore deposits. Thus the Ganeshwar-Jodhpura-O.C.P. complex seem to have used the ore resources in their immediate vicinity. Its confirmity is due to the fact that in north-east Rajasthan following sites are reported from where copper slags are present *in situ*. These are: 1) Jambalpur, 2) Chinchroil, 3) Oda-ki-Dhani, 4) Aheerwala, 5) Basedi, 6) Bihar, 7) Chamrauli, 8) Delsar, 9) Dhooniwala, 10) Dilpura, 11) Balwara, 12) Gaonri, 13) Goylari, 14) Khalra, 15) Kulra, 16) Kishore Para, 17) Loharwas Sota, 18) Nimodh, 19) Para Bhitri ki Dhani, 20) Purana Mothuka, 21) Pyojara, 22) Tigri, 23) Umrawala, and 24) Khetri. Apart from this Khetri and Bahar have copper ore deposit. The South-East Rajasthan is dominated by Ahar culture. Over fifty sites of this culture are distributed in the basin of the Banas and its tributary system comprising the Berach,, the Kothari, the Khari, the Dai, the Sodra and the Mashi and their smaller tributaries in the districts of Udaipur, Chitorgarh, Bhilwara, Tonk, Ajmer and Jaipur in Rajasthan and Mandasor in Madhya Pradesh. Ahar itself yielded heaps of copper slag whose spectroscopic analysis confirm that they came from chalcopyrite ore from the Khetri mine in North-East Rajasthan.[46] But local oxide ore of the copper was also used by Aharian people was tested and confirmed by the author himself. The presence of Ahar level at Kayatha confirm their orientation towards the Malwa region. Here Malwa includes all lands between the Vindhyas and a point just south of Gwalior, its eastern limit is marked by a ridge which goes from south to north near Vidisha, while its western limit lies near the Rajasthan border. In Malwa, copper ores are to be found in Jabalpur in the form of malachite and oxide ore on a bed of Bijawara quartize in the neighbourhood of Sleemanabad. A strip of 29 kms long and 5 kms broad around Sleemanabad and Hiwar have copper ore deposit. Definitive information on this copper ore belt known as the Sidhi-Shahdol-Jabalpur belt is available in the records of G.S.E.[47] A major portion of the copper belt which is N 60 E – S 60W (250 km long) is under an alluvial cover.

In south of Malwa, Maharashtra plateau is present which is geographically a region of homogeneity by means at a variable altitudes and the wide open valley in which the present rivers appears like misfit streams.

The Neolithic-Chalcolithic cultural sequence of the plateau is perhaps the most scientifically dated in Indian archaeology under the auspices of the Deccan College, Pune. Nevasa, Jorwe, Nasik, Chandoli, Songaon and Inamgaon, among other sites, have been excavated by themselves. The Archaeological Survey of India has also taken interest in the region and excavated Bahal, Prakash, Daimabad, Tuljapur Garhi etc. The presence of local copper smelting is reported only at Daimabad (19°31'N; 74°42'E) and Inamgaon (18° 35'N; 74°70'E). At Daimabad in phase I a small portion of a furnace which also yielded a slag is reported by Sali.[48] According to him, the copper-smelting furnace referred by Rao, in previous excavation at the same site, also belonged to the Daimabad cultural phase IV yielded in House No. 9 the remains of two furnaces, both of which contained ash, while one also yielded a heart-shaped razor of copper and this house has been referred as the coppersmith's workshop. It yielded a crucible too which was used by metal-craftsman. Furthermore, slag has been encountered in the Harappan phase. At Imamgaon (18° 35', 74° 30') the evidence consists of remains of a furnace in the middle levels of Period II, early Jorwe culture (1400-1000 B.C.) and a lump of copper ore from the late level of Period III[49,] late Jorme culture (1000-700 B.C.). In Maharashtra, Nagpur, Chandrapur, Bhandara and Ratnagiri districts have copper ore zone. In Nagpur at pular-Passore a 3 kms length of copper ore zone of brecciated by quartz veins is present while at Dangarmunda in the same district, disseminations of chalcopyrite and malachite encrustaitions occur in the quartz zones traversing the breccia zones which is a fissure containing numerous wall-rock fragments, with mineral deposits in the interstices. In Chandrapur district at Thanewasna, the mineralization zone strike a length of 2.2 kms,. is made up of chalcopyrite, arsenopyrite and pyrite. At Dubharpeth, the surface evidence of mineralization are seen over a 5 km strite length, while Ghota zone is 6 km long. At Govindpur in the same district, copper mineralization is noted along the shears and joints in the trap rocks. In Bhandara district, disseminations of chalcopyrite and pyrite have been noted at Mendaki, while in Ratnagiri district at Hedavada, disseminations of chalcopyrite associated with a small awdent of bromite and pyrite were found in the quartz veins. Native copper was also recorded at Kolhapur, Jalakwadi, Rasin and Sirsondi. Dhavalikar[50] noted native copper at Rashin and Sisur in Ahamednagar district, not far from Imamgaon.

Thus during chalcolithic period, Maharashtra had significant contact with central India from the late Harappan through the Malwa phase and acted as a corridor along with mineral resources from Gujarat and cultural elements percolated through to that region. With Gujarat also, Maharashtra shared an intimate relationship as manifested in the presence of late Harappan at a number of sites by the presence of Lustrous Red Ware in the sub-region stretching from Kaothe (District Dhule) near western coast to Tuljapur Garhi in the Amrawati district. By the Jorwe period (1400-700 B.C.), however, Maharashtra became more southward orieinted with the Jorwe ceramics found in Andhra Pradesh and southern Mysore in karnataka.

In Andhra Pradesh a number of copper old working are found in the Kurnool, Guntur, Mahbubnagar and Prakasam districts. In the Kurnool region, these include working at irregular intervals in an area stretching over 16 km in the Gani-Kalava valley, shallow pits near Kalava, Gani, Chennakkapalle and Brahmanpalle.[51] In the Guntur district, the ancient copper working around Agnigundala have been referred to since the beginning of the nineteenth century.[52] Huge slag heaps around the villages of Agnigundala, Ipuru, Vadujerala, Malapadu, and Yamagiripaya and some places broken earthen tubes, believed to be broken crucibles for smelting the ore, were also noted. Ancient working for copper are found at few places in the Makthal Taluk in Mahbubnagar district of Andhra Pradesh.[53] Therefore, the chalcolithic phase of Andhra Pradesh have been basically a regional phenomenon, used locally available copper ore resources.

In Karnataka ancient workings for copper ore have been noticed at western flanks of Belligudda hill near Ingladahalle in Chitradurga[54] district, Tinthin in Gulbarga district, Kaiga area in North Kanara district, Kappatgudda hill in Dharwar district[55], Mandargi, Chikhesuru and Hirehesura in Riachur district[56] and Khajjidoni in Bijapur district.[57] On the basis of material culture of the chalcolithic phase, one important conclusion can be made that although contact of some significance is evident between Maharashtra and north Karnataka, there is not enough ground to claim that this contact was sufficiently intensive to suggest an actual large-scale migration from Karnataka through the Krishna Valley into Maharashtra. An assessment of the resources utilized and available in this sub-region of northern Karnataka, however, clearly shows that no raw material need too have been produced from Maharashtra. Like northern Karnataka, the southern uplands of the State also seem to have been sharing in the general interaction with its northern counterpart, but the copper ore is present in very small quantities. By this fact, the north Karnataka have large number of chalcolithic sites in contrast to south Karnataka.

The Ganga plain is a thick pile of the Quaternary deposits, whose width in the north-eastern India is only 70-85 km. But widens to a maximum of 400 km. in Bengal; while in western Uttar Pradesh it is of the order of 150-300 km. The maximum depth of the alluvium up to the basement rocks is about 6 km based on computations of seismic sounding. Its basement has an uneven topography and that the thickness of the sedimentary deposits is highly variable. Seismic and gravity studies have indicated that the crystal thickness beneath the Gangetic plains is 28-45 km, including a 3 km top sedimentary layer. The sedimentary sequence rests uncomformably upon crystaline basement, lying 2,000 m to over 3,000 m below the surface. Underlying the alluvium in the Ganga Basin are less consolidated Shiwalik and older Tertiary sediments of the Himalayan piedmont, and lying unconformably below these are more consolidated, older formations referable to the Vindhyan Supergroup. In some sections sediments such as Gondwana and the Cretaceous may also be present.

Archaeologically, the Upper Gangetic plains which lies approximately between the Yamuna in the West and 100 m contour in the east (73° 3'E - 82° 21'E and 25° 15'N - 30° 17'N) has been dominated by ochre ware pottery people along with late Harappans. It has been argued that O.C.P. people were the original inhabitants of the region and that with the advent of the

Harappans, a part of the population moved east as far as central Doab. On the basis of thermoluminescence dating from Atranjikhera (27° 42'; 78° 44'), Lal Qila (28° 46'; 77° 15'), Jhinjhina (29° 12';77° 18') and Nasirpur (29° 46'; 77° 52'), B.B. Lal suggested that the OCP/copper hoard culture was in existence between 2000-1500 B.C.E. in the Upper Gangetic plain. The subsequent cultural phase was marked by plain black-and-red ware. The Upper Gangetic plains are a vast alluvial expanse cover an area of 149029 sq.kms. with no significant occurrence of the basic raw materials that were used within the sub-region in this period. The only exceptions are limited occurrences of copper in its fringes. A reasonably exhaustive account of copper ores and old workings in western Uttar Pradesh and Uttaranchal have been discussed in details in the Bulletins of the Geological Survey of India No. 47. In Kumaun district copper is noted from Rai mine, Sira mine and the Gaul mine. Gani, Phadiall, Giwar, Belar, Sor are the important sites in Kumaun district where old copper working was reported.[58] In Garhwal district the copper mines are reported from Dhanpur, Dhobri, Agar, Sera and Pokhri[59], while in Pithoragarh district copper deposits have been found at Askot, Bora, Agar, Rain Agar, Khanayan-Galpakot-Kimkhet area and Kanalichinna.[60] In Almora district at Bageshwar, a heavy deposit of malachite is reported.[61] The hills in Biana and Wair in Bhartapur district of north-east Rajasthan and Antri, Khrol and Bhagesh-Gijoha copper deposit in Gwalior district of Madhya Pradesh were also important copper fielders to Upper Gangetic chalcolithic people. Thus the Upper Gangetic plains were involved in interaction with the Uttaranchal, Baghelkhand area of Madhya Pradesh and North-East Rajasthan so far as copper ores are concerned during its chalcolithic phase.

The middle Gangetic plain (82° 21' - 87° 16'E and 25° 15'N - 27° 45'N) cover an area of 99200 sq. kilometers play a transitional zone between the upper Gangetic plains and the deltaic zone of lower Gangetic plains. Culturally in the period of chalcolithic, we see different cultural elements within the sub-region in the form of Black-and-Red Ware phase of Chirand (25° 45'; 80° 50') is represented at other sites in Gaya, Vaishali, Bhagalpur in Bihar and Basti, Deoria, Ballia, Mirzapur and Varanasi in eastern Uttar Pradesh. Secondly, we have the sequence of the Sarayupara region of north-eastern Uttar Pradesh, which is fairly widespread, extending over a large part of Ghaghara-Rapti interplume. Thirdly, in Allahabad district at Sringaverpur whose red ware phase of Period I is significantly different from the rest of the region. The source of copper for this region must be Singhhhum, Hazaribagh and Santhal Parganas of Jharkhand and Bhagalpur in Bihar were heavy deposit of copper is present. In Singhbhum a belt stretches for 128 kms between Duarpuram in the west and Baharagora in the south-east have evidence of ancient mine working and copper slag heaps are present all over the belt. This copper belt of Singhbhum overthrust in the Dharwar schists and intrusive granite. The deposits worked at the Mosaboni mine consists of sulphide ore assaying 2.4 per cent of copper.

The metallographic and chemical study of Chalcolithic copper artifacts from Senuwar (24° 56'N; 83° 56'E) in Rohtas District of Bihar in middle Ganga plain confirm that they obtained copper ores from Rakha mines due to the common presence of bismuth, gold, zirconium, tungston and titanium as trace elements. At the same time arsenic and tin do not coincide with each other. The present knowledge, however, is inadequate regarding early mining activities in this area in early phase of 1800 B.C.E. Besides definite information on the oxidised ores are also not available. However, there is some chance for getting copper ore from Rakha mines but definitely not from chalcopyrite of Singhbhum. But the metallographic examination of copper artifacts from Manjhi in Saran District of Bihar confirm that during 400 B.C.E. to 300 A.D. the copper was produced from the chalcopyrite sulphide ore of Singhbhum of Jharkhand.

The lower Ganga plain (87° 16' - 88° 45'E and 22° 10' - 25° 15'N) is the vast deltaic zone lying to the south of the Rajmahal-Garo alignment and embracing an area of about 80,968 sq. kilometers from Bhagalpur to Bay of Bengal. The chalcolithic phase of this area obtained copper either from Jharkhand or from Purulia-Bankura shear zone, with some minor occurrence of copper ores in Midnapur and Bankura districts of West Bengal. In the

Bankura district two old copper mines was noticed at Chhedapather and Tamkum.[62]

Chemical Analysis of Chalcolithic Metal artifacts

S.No.	Site	Cu	Sn	Zn	As	Pb	Sb	Ni	S	Fe
1.	Mohenjodaro	99.67	-	-	0.15	0.02	0.88	1.28	0.98	0.03
2.	- do -	97.07	-	-	0.98	Tr	Tr	0.31	1.15	0.49
3.	- do -	96.42	-	-	-	0.09	-	0.37	0.36	-
4.	- do -	92.49	0.37	-	1.30	Tr	Tr	1.06	2.26	1.51
5.	- do	95.80		-	0.74	1.58	0.72	0.25	0.61	0.12
6.	- do -	94.76	0.09	-	4.42	0.26	-	0.14	-	0.15
8.	- do -	92.41	-	-	3.42	3.28	0.10	0.15	0.05	0.59
9.	- do -	88.05	8.22	-	Tr	2.60	-	Tr	0.84	0.29
10	- do -	86.22	12.38	-	-	0.70	0.35	-	-	0.35
11	- do -	82.71	13.21	-	1.17	0.56	0.33	0.56	-	0.42
12	- do -	85.37	11.09	-	0.07	Tr	Tr	0.16	0.11	0.18
13	- do -	83.92	12.13	-	-	0.17	Tr	0.17	-	-
14	Harappa	97.66	0.33	-	0.06	0.70	-	0.14	1.02	1.11
15	- do -	98.12	0.33	-	0.65	0.10	-	0.36	-	0.41
16	- do -	98.60	0.07	-	0.66	tr	-	0.26	0.21	0.41
17	- do -	98.96	0.10	-	0.68	tr	-	0.40	-	0.13
18	- do -	97.69	0.15	-	1.19	0.85	-	0.05	-	0.07
19	- do -	98.37	-	-	1.40	0.11	-	0.10	-	0.02
20	- do -	92.35	0.29	0.07	2.96	3.72	-	0.21	-	0.20
21	- do -	90.05	-	-	6.58	2.80	-	0.18	0.21	0.39
22	- do -	97.20	0.84	-	0.70	-	0.19	0.09	-	.098

S.No.	Site	Cu	Sn	Zn	As	Pb	Sb	Ni	S	Fe
23	- do -	94.92	3.60	-	0.60	0.20	-	0.29	0.91	0.39
24	- do -	92.61	6.43	-	0.36	tr	0.31	0.20	-	0.90
25	- do -	91.87	6.42	-	0.26	0.98	-	tr	-	0.47
26	- do -	91.00	6.76	-	0.04	0.04	0.44	0.14		0.74
27	- do -	91.10	7.85	-	0.42	tr	tr	0.22		0.41
28	- do -	88.38	9.16	-	0.40	0.10	0.42	0.17		1.37
29	- do -	87.42	10.45	-	1.10	0.52	-	0.17	0.14	0.34
30	Ganeshwar	97	0.01		0.30	1.00		0.60		
31	- do -	96.5	0.2	0.25	1.00	0.03		0.04	-	0.2
32	Ahar	98.65	7.74	0.296	2.06	0.53	0.24	0.59		0.257
33	Navadatoli	99.91	10.4	0.479	0.32	0.67	0.35	0.23		0.476
34	Chandoli	99.90	0.425	0.445	0.442	0.689	0.34	.404		0.443
35	Jorwe	99.11	0.135	0.496	0.681	0.365	0.42	0.179		0.402
36	Nevasa	96.9	0.198	0.501	0.13	1.18	.445	0.38		0.289
37	Kayatha	95.56	0.58	0.498	0.24	1.22	0.213	1.12		1.36
38	Inamgaon	98.23	1.81	0.514	0.184	6.49	0.03	0.59		0.08
39	- do -	99.12	1.19	.494	0.53	0.463	0.45	0.1		0.0125
40	- do -	95.45	6.14	.394	4.84	0.707	0.37	3.50		0.0125
41	Senuwar	94.65	6.42	0.95	1.05	0.005	0.10	0.05		0.04
42	Manjhi	92.46	8.80	0.17	-	1.45	-	0.01	-	0.07

Bronze is a copper based alloy of copper and tin, other than a brass which is basically a copper-zinc alloys containing more than 50% copper and atleast 40% zinc. Some time brasses contain 30% zinc, but this 30% zinc containing brasses are less harder than 40% zinc containing brasses. In an independent condition copper is much harder than tin, but in the cases of bronzes, tin hardens and strengthens the copper. Since the melting point of tin is 232^{0}C and copper is 1083.4^{0}C thus tin can be easily obtained from its ores so it is known since from prehistoric times. It is a soft, brilliant white metal with corrosion-resistance feature, but its solubility in copper rapidly decreases below 350^{0}C (becoming approximately 1% at normal room temperature). Between 350^{0}C to 500^{0}C transformation of tin into copper is very sluggish but Cu_3Sn structure only appear after long annealing treatment by which bronze become as harder as iron. The annealing heat treatment promote homogeneity in the structure of copper-tin alloy by diffusion, relieve residual stresses arising from cold work operations and may also result in grain refinement by recrystallization. The temperature to which the metal is raised and the time for which the temperature is maintained depend upon both metal concerned and the particular effects that are desired. The aforesaid carried out chemical analysis of chalcolithic artifacts confirm that Harppans were much acquainted with tin alloying with copper from 4% to 12% which is not seen in other chalcolithic culture of India. But arsenic bronzes is a salient feature of all the chalcolithic artifacts. Arsenic is used in copper to impart oxidation-resistance and a measure of high temperature strength. The above analysis also depict that almost all the chalcolithic artifacts copper is alloyed with arsenic and lead. Here lead in alloys always to promote hardness and minimise segregation.

The antiquity of zinc goes back to late Harppan phase, but typical brass is absent. At Lothal (22° 31'; 72° 15') a copper object of 1500 B.C.E. have 6.04% zinc and at Rojdi (21^0 51'; 70^0 54') some chisel, celt, rod and bangles containing upto 1.54% zinc. A part of a chariot is submerged. Dwarka (22° 15'; 69° 0') of *circa* 1500 B.C.E. assays 10.68% zinc, 1.32% lead, 0.43% iron and rest of copper. Thus it is also to be noted that these early samples of cementation brass which did not contain the typical brass having atleast 40% zinc is due to ore impurity which can be only seen in the region of Gujarat and Rajasthan but not other part of India during *circa* 1500 B.C.E. These early samples of zinc containing copper by its ore impurity confirm their possible link with Rajpura-Dariba silver-lead-zinc ore mine near Udaipur having 1260 ± 160, 1136 ± 160; and 1050 ± 150 B.C.E., C-14 dates of the timber sample. In the same district of Udaipur Zawar lead-zinc deposit is the most famous, not merely for its abundant traces of old workings extended all over 25 km mining belt and apparently going down to a depth of 90 meter below surface. More than 1 km valley leading to the mining area of Zawar is covered by smelting debris and clay retorts for zinc smelting. The C-14 date at Zawar mines goes back to 755-685 BCE of wooden sample from scaffold but their maximum frequency of dates are in the range of Early Medieval Period (700-1200 A.D.) At Dariba in District Rajsamand and Ajucha in District Bhilwara of Rajasthan, mine workings, heaps of mined out debries, slag, furnace refractories and large timber revetements are obtained by Hindustan Zinc Limited whose C-14 date goes back to *circa* 1000 B.C.E. This was the sharp edge of technical innovations of zinc, taking place on a major scale, well away from West Asia, Europe and Mediterranean in Rajasthan and Gujarat. In this region the brass was made through the process of cementation for diffusing elements into a material through the surface. In this process the reduction of copper and zinc ore in the same furnace simultaneously taking place in which the zinc oxide is easily reducible at usual temperature of the smelting furnace, the metal zinc was produced in the vapour state, since it has low boiling point 917° C. Zinc vapour absorbed in copper to produce cementation brass. But zinc extraction from its ore is indeed a very difficult process. It oxidises very rapidly and has to be distilled in an air tight reducing atmosphere at 1100°C. By this complication its extraction was very late in Indian metallurgy. Again during P.GW. phase of Atranjikhera (27° 42'; 78° 44') produced heavily alloyed copper objects containing 6.28 to 16.20% zinc with 11.68 to 20.72% tin. But all the above mentioned zinc alloy of copper is not typical brasses. The typical brasses are of two main types of alloys: (a) the α-brasses contain up to 30% zinc and (b) the β-brasses contain about 40% zinc. These alloys of brasses have excellent corrosion-resistance, good mechanical properties and good drowing properties. They are harder and stronger and require a very hot working and extrusion.

Notes and References

1. The thickness of the earth crust is from 10 km (beneath the oceans) to 50 km (beneath the continents) which form the lithosphere; outside of the hydrosphere of the earth.
2. The igneous rocks form when molten magma cools and crystallizes. They constitute about 15% of the earth's surface, and are formed of molten material (magma) that flows up from the deeper part of the crust.
3. The sedimentary rock, formed from pre-existing rocks by the erosion of the latter into particles of varying sizes, which are subsequently transported and deposited as sediments usually in seas and lakhs. As these sediments accumulate they form sedimentary rocks.
4. The metamorphic rock formed on pre-existing solid rocks by mineralogical, structural and chemical changes in response to extreme changes in pressure, temperature and shearing stress constituent minerals have been recrystallized and reoriented and new minerals grown without the whole rock melting; may be a result of contact, dynamic, or regional metamorphism. Metamorphic rocks are significant because they constitute a large part of the continental crust.
5. Vein is a general term for a crack or fissure in the earth's crust in which highly heated water from below have deposited crystalline minerals (especially vein quartz) from solution, and under special circumstances, metallic minerals of economic importance.
6. Bedded deposit is the arrangement of rock strata in bands of various thickness and character.
7. Phyllite is a medium grade regional metamorphic rock of light silver-grey colour with a texture of grading from granoblastic (quartzone phyllities) to prophyroblastic (albitic phyllite) derived from clayey sedimentary rocks containing a residue of organic material plus chlorite and mica.
8. Schist is a regionally metamorphosed rock characterized by the presence of quartz, feldspar, mica, chlorite, talc or hornblende, schistone and lineated texture.

9. Slate is a group name of various fine grained rocks derived from mudstone, siltstone and other clayey sediments as a result of low degree regional metamorhism. It develops under temperatures and pressure only slightly greater than those found in the sedimentary realm. The temperatures are not high enough for the rock to crystallize. The important controlling factor is intense shearing or compressive direct pressure.
10. The Delhi-Aravalli system of strata is probably of Lower Cuddapah age, though in its form the general tectonic features of this system. It occupies a large extent of Eastern Rajasthan extending from Delhi to Idar in constricted, sorely eroded synclinal bands in the center of the great Aravalli synclinorium, its fullest development being found in the main Rajasthan geosyncline of Ajmer-Marwara and Mewar area.
11. Zaskar river originated from Indus in Jammu and Kashmir and move towards south-east through Zaskar mountains and merged into north of Manali in Himanchal Pradesh.
12. Kenoyer, J.M. (1991), "The Indus Valley Tradition of Pakistan and Western India," *Journal of World Prehistory*, 4(5), pp. 331-385.
13. Stein, Sir Aurel (1929), "An Archaeological Tour in Waziristan and Northern Baluchistan," *Memoirs of the Archaeological Survey of India, No. 37*, Calcutta.
14. Hargreaves, H. (1929), "Excavation in Baluchistan, 1925: Sampur Mound, Mastung and Sohr Damb, Nal." *Memoirs of the Archaeological Survey of India No. 35*, Calcutta.
15. De Cardi, B. (1965), "Excavations and reconnaissance in Kalat, West Pakistan: The Prehistoric Sequence in the Surab region," *Pakistan Archaeology* No.2, pp. 86-182 (Karachi).
16. Fairservis, W.A.,Jr. (1956), "Excavations in Quelta Valley, West Pakistan, 2 pts." *Anthropological Papers of the American Museum of Natural History*, Vol. 47, (New York).
17. Dales, G.F. (1979), "The Balakot Project: Summary of four years excavation in Pakistan," in *South Asian Archaeology 1977*, Vol. I, pp. 241-74.
18. Casal, J.M. (1961), *Follios de Memdigak*, 2 Vols. (Paris).
19. Jarrige, J.F. and M. Lechevallier (1979). "Excavation at Mehargarh, Baluchistan: Their significance in the prehistorical context of the Indo-Pakistan borderlands." In *South Asian Archaeology*, 1977, Vol. 1, pp. 463-535.

20. Jarrige, J.F. and R.H. Meadow (1980). "The antecedents of Civilization in the Indus Valley," *Scientific Americana*, August (243, No. 2), pp.122-33.

21. The time of Period VII at Mehrgarh is middle of the third millennium B.C., which spread over the main mound and their material culture are parallels to Damb Sadat III; Mundigak IV; Shaha-i-Sokuta II; Kot Diji; Amri II B.

22. Habib, I. (1982). *An Atlas of Mughal India* (Delhi), notes to sheet 5b.

23. Minchin, C.F. (1907d). *Baluchistan District Gazetteer Series VIII Las Bela* (Allahabad), pp. 118-119.

24. Hoste, De la (1840) "Memoranda Representing the Existence of Copper in the Territory of Luz." *Journal of Asiatic Society of Bengal*, IX, pp. 30-33.

25. Minchin, C.F. (1907c), *Baluchistan District Gazetteer Series VI VIA and VIB. Sarawan, Kachhi and Jhalawan*. Bombay, p. 161.

26. Buller, R.H. (1906). *Baluchistan District Gazetteer Series IV and IVA Bolan and Chagai*" (Karachi), p. 107.

27. Hasan, M.U. (1986). "Recent Archaeological Finds in Baluchistan and Their Implications," *South Asian Archaeology Congress* (Delhi).

28. Dales, G.F. (1962). "Harappan outposts on the Makran Coast," *Antiquity*, No. 36 (Cambridge), pp. 86-92.

29. ______ (1979). "The Balakot Project: Summary of Four Years Excavation in Pakistan" in *South Asian Archaeology*, 1977, Vol. 1, pp. 241-74.

30. Mughal, M. Rafique (1973). *The Present State of Research on the Indus Civilization*. Karachi, Department of Archaeology, Government of Pakistan.

31. Ball, V. (1881). *A Manual of the Geology of India*, Part III. Economic Geology (Calcutta), p. 226.

32. Oldham, R.D. (1893). *A Manual of Geology of India Stratigraphical and Structural Geology* (Calcutta), p. 489.

33. Agrawal, R.C. & Kumar, V. (1982). "Ganeshwar-Jodhpura Culture: New Traits in Indian Archaeology." *Harappan Civilization* (G.L. Posshel ed.) (New Delhi), pp. 125-34.

34. Sanaullah, M.K.B. "Sources, composition and Technique of Copper and its Alloys," in *Excavation at Harappa* by M.S.Vats, 1 (Delhi, 1940), pp. 378-383.

35. Agrawal, D.P. (1970). "Metal Technology of Harappan Culture and its Socio-Economic Implications," *Indian Journal of History of Science*, Vol. 5 (2), pp. 238-252.

36. Agrawal, R.C. (1984). "Aravalli, the Major source of copper for the Indus and Indus related Cultures," *Frontiers of the Indus Civilization* (ed. B.B. Lal and S.P. Gupta) (Delhi), pp. 156-62.

37. Raghu Nandan, K.R. Dhruva Rao & M.L. Singhal (1981), *Exploration for Copper, lead and Zinc Ores of india*, Bulletin of the G.S.E. Series A—Economic Geology (Calcutta), pp. 57-61.

38. The copper ore deposits in Zhob, Robat Shah Bellaul area and between Ziarat (Quetta district) and Sohan (Sarwan district) of Baluchistan.

39. Mughal, M. Rafique (1981), "New Archaeological Evidence from Bahawalpur." *Indus Civilization—New Perspective*, ed. A.H. Dani (Islamabad), pp. 33-41.

40. Gupta, S.P. (1996), *The Indus Saraswati Civilization* (Delhi).

41. Mughal, M. Rafique (1982). "Recent Archaeological Research in Cholistan Desert," *Harappan Civilization* (ed. G.L. Possehl) (New Delhi), pp. 85-96.

42. Antonini C.S. and Stacul, G. (1972). *The Protohistoric Graveyards of Swat (Pakistan)* (Rome), pp. 15-20.

43. Dani, A.H. (1988). *Recent Archaeological Discoveries in Pakistan* (France, Tokyo), pp. 120-24.

44. Coggin Brown, J. and Dey, A.K. (1955), *India's Mineral Wealth* (London), p. 153.

45. Varadan, V.K.S. (1976a) *Geology and Mineral Resources of States of India. Jammu and Kashmir State G.S.I. Miscellaneous Publication No. 30* (Calcutta), pp. 44-46.

46. Hedge, K.T.M. (1969). Technical Studies in Copper Artifacts from Thar, *Excavation at Ahar* (H.D. Sankalia, S.B. Deo & Z.A. Ansari) (Poona), pp. 225-228.

47. *Bulletin of the G.S.I. Series A. Economic Geology No. 47* (Calcutta, 1981), p. 39.

48. Sali, S.A. (1986). *Daimabad MASI No. 83* (New Delhi), pp. 93-98.

49. Dhavalikar, M.K. (1976). "Settlement Pattern of Imamgaon", *Puratattva* No. 8, pp. 44-52.

50. *Ibid* (1976), p. 52.

51. Satyanarayana Rao, P. (1072) *Bulletins of the G.S.I. No. 36* (Delhi), pp. 26-27.

52. Sivadas, K.M. (1985). "Lead and Copper Deposits of the Agnigundala Mineralized Belt," *M.G.S.I.* 118 (Calcutta), p. 1.
53. Rajgopal, M.V. (1976a). *Andhra Pradesh District Gazetteers Mahbubnagar* (Hyderabad), p. 9.
54. Karunakaran,C.K. (1974a). *Geology and Mineral Resources of the State of India*, Part VII (No. 30) (Delhi), p. 13.
55. Palandey, M.R. (1959). *Gazetteer of Bombay State Dharwar District* (Bombay), p. 16.
56. Abhishankar, K. (1970). *Maysore State Gazetteer, Raichur District*. (Bangalore), p. 15.
57. Sri Sathyan, B.N. (1966b). *Maysore State Gazetteer, Bijapur District* (Bangalore), p. 23.
58. Ball, V. (1881). *A Manual of the Geology of India*, Part III (Calcutta), p. 269.
59. *Ibid.*, p. 270-272.
60. Raghunandan (1981), *Economic Geology* No. 47 (Calcutta), p. 190-195.
61. *Ibid.*, p. 195.
62. Chakraborti, D.K. (1985-86). "The Pre-Industrial Mines of India." *Puratattva*, No 16, p. 67.

CHAPTER 4

FERROUS METALLURGY

Ferrous is the traditional name identifying the lower common oxidation number of iron, now designated by iron^{+2} and ferro is the prefix indicating an alloy (or compound) of iron or steel.

After silicon which have 28% by weight of the earth's crust, iron is the second most metal i.e. 5% by weight of the earth's crust of Precambrian age[1], whereas copper is only 0.5% by weight of the earth's crust. Although iron is much more abundant than copper on the earth but its exploitation comes later in history and its technology follows a vastly different course, for reasons inherent in the metal itself. For more than any other metal in common use, iron affected in its properties by the techniques used to produce and work it, from the smelting process (which may add alloying carbon which is a non-metal, with metallic iron) to the final shaping, hammering or pyrotreatment. The exploitation of any metal naturally depends on the cost of producing it and the qualities it can be seen to offer; in the case of iron, however, the most useful properties of the metal are not readily apparent and were only slowly evoked by the ingenuity of generations of craftsmen.

All the world's important iron deposits are sedimentary in nature. During chemical and mechanical weathering, the materials from the source of rocks are liberated and carried out either in suspension or in solution, by natural agencies like streams to the depositional sites. The deposition of the particles in suspension (mechanical load) depends on factors such as the velocity, gradient of channel, size and shape of the particles and their specific gravities.[2] These deposits occur as alluvial placers in the earth's crust. India has around 60% of the world's high grade iron ore reserves. The total reserves of iron ore with 55% and above of iron are estimated to be 9,050 million tones. Of the three types of deposits i.e. magnetite (Fe_3O_4), hematite (Fe_2O_3) and limonite ($2Fe_2O_3.3H_2O$), the hematitic variety which is red in colour consisting about 80% of the total reserves is of considerable importance because of its high grade quality. The iron content of hematite is very high—from 55% to 65%, while that of harmful impurities like sulphur and phosphorus is very low. It is more difficult to reduce magnetite than other iron ores due to its high content sulphur (upto 1.5% to 2%) and same problem is with limonite which contains high percentage of phosphorus. The gangue[3] in hematite is most often quartzite, while in magnetite is of a siliceous nature with an admixture of certain amount of other oxides. The cherry red colour of hematite and black colour of magnetite can be correlated with *Lohit* and *Shyam ayas* of Vedic text. Here *Lohit* means red and *Shyam* means black.[4]

About 43% of the Indian iron ore reserves are situated in Jharkhand and Orissa and neighbouring part of West Bengal. In Peninsula, where the crystalline and schistose rocks of the Dharwar and Cuddapah system enclose at some places ferruginous deposits of an extraordinary magnitude is present. Among these, massive outcrops of hematite and magnetite of the dimensions of whole hills are not unknown. But the most common mode of occurrence of iron is as laminated hematite, micaceous hematite and hematite-breccia; lateritic hematite also forms large deposits, together with hematite-and magnetite-quartz-schists, the metamorphosed products of original ferruginous sands and clays. The high-grade hematite ore-bodies of Singhbhum in Jharkhand, together with those of Bastar in Chhattisgarh,

Keonjhar, Bonai and Mayurbhanj of Orissa, Karnataka, Tamil Nadu, Goa are of Upper Dharwar or newer age, the remarkable concentration of the iron ore in them being ascribed to post-Guddapah metasomatic action[5], to original marine chemical precipitation of oxides, carbonates and other compounds of iron, to volcanic action and other agencies. These ore-bodies[6], many of them containing 60-65% of iron, are thought to be the largest and richest deposit of iron perhaps in the world. The Damuda series of West Bengal which spread in the area of Jharia and Raniganj holds valuable deposits of beaded or precipitated iron ore in the iron stone shades. In Deccan traps iron is a prominent constituent of laterite.[7] Laterite is a Greek word means brick red coloured formed in the tropical region of South India and Africa. In Indian context this term was first used by F. Buchanan in 1807 from Malabar hill of Kerala. It is a type of sesqui oxide of iron (in red colour) and alumina (in grey colour) containing 40% to 50% iron formed during Pleistocene phase under alkaline weathering of Dharwar and Khondalite. In Chhattisgarh the deposits occur in Bailadila range of Bastar and Dalli-Rajhara hills of Durga district. The magnetite type of iron ore in India constituting 8% of the total reserves is either of igneous origin or metamorphosed banded-magnetite silica formations, possibly of sedimentary origin. The apatite-magnetite formations is seen in Singhbhum in Jharkhand shear zone and the titaniferous[8] and vanadiferous[9] magnetite in Mayurbhunj district of Orissa. The banded magnetite silica deposits occur mainly in Salem and Trichirapalli district of Tamil Nadu (with 35% iron), the Kudermukh deposit of Karnataka (with 43% iron) and in Guntur district of Andhra Pradesh (with 33-37% iron content).

A vast corpus of work has grown up around many of the issues in connection with the beginning of iron technology and its archaeological and cultural ramifications, the time and space of different ware types such as the Painted Grey Ware, Black-and-Red Ware and the Northern Black Polished Ware and the cultural antecedents of the Megaliths. In addition to the archaeological data, there becomes available, for the first time, a body of literature: Brahmanical, Buddhist, Jain and Greek accounts, which, in spite of the later interpolations, substantially reflects the socio-economic conditions of the iron age in India.

The appearance of iron in Kashmir coincides with the historical period which is characterised mainly by the presence of the Northern Black Polished Ware (NBPW) and associated Red Ware (RW). On the basis of W.A. Lawrence's work[10], Chakraborti[11] has noted a tradition of pre-industrial iron smelting in the region of Kashmir. Lawrence noted that iron has been discovered and worked in several places in the valley, and there are extensive workings of iron ore in the neighbourhood of Sof. The Kashmiries regard the Sof iron as superior to the iron imported from India for the purposes of agricultural implements, and the blacksmiths always speak of it most favourable. This iron ore was procured from a bed of impure calcareous limonite ($2Fe_2O_3.3H_2O$), found in the rocks east of Achabal, Sope, Ouri and Khrew.[12] Since limonite is a weathering product of iron containing minerals and is frequently intermixed with many other substances including colloidal silica, clay minerals, aluminium hydroxide etc., thus may be considered as a minor ore of iron and also known as brown hematite. Workable iron-ore is met with in the Riasi district, Jammu hills, in association with the Nummulitic series, which supported a number of local furnaces for the manufacture of mutinious of war during the last two centuries in Jammu and Kashmir.

The affinity of the Kashmir valley with the North West Frontier Province is evident in, among other things the presence of an Early historic red ware associated with NBPW and iron at Gufkral III and Semthan II of Kashmir. At the head of the doab of the Indus and Jhelum rivers, the Potwar plateau was dominated by one of the most flourishing Early historic centres of Ancient India, as informed by the excavations of Taxila, Manikyala, Hund, Thareli, Chanaka Dheri, Udegram, Butkara, Nimogram, Chakadra, Chasada and Shabajgarhi in Pakistan. Muhammad Sharif has discovered abundant iron slag from the Period IV of the cultural debris of Taxila.[13] Low grade ferruginous laterites in Potwar range were known to have been worked for small-scale local smelting. Coggin Brown and Dey[14] noted a survey made in 1952 that the Mianwali, Sargodha and Attock districts of Punjab in Pakistan have workable deposits of iron ore containing 40% iron and at Fatehganj and at Hasan Andal in Attock district he noted a long tradition of iron smelting and prepare padlocks and stirrups of iron. Thus an overview of the ore materials available within the Potwar plateau and its immediately adjacent regions, suggests that many of the basic iron ore resources that the city-state like Taxila demanded could be found in the ambit of the sub-region and the North West Frontier province itself.

The North West Frontier Province of Pakistan constituted the Gandhara of Indian historical tradition took a different course of Post-Harappan cultural development, the most important discovery was made in Peshawar Valley, Swat, Dir and Bajaur. This culture is related to a type of grave, which was first discovered by the Italians[15] in Swat and termed as "pre-Buddhist". Later Dani[16] located similar graves in the southern part of the Dir district, Bajaur, and in the Peshawar valley north of the Kabul river, and rechristened them the "Gandhar Grave Culture", because they were first discovered in Gandhara and covered almost whole of this ancient region. The Italian[17] have now renamed them "Protohistoric graves". The third-period graves of this culture have for the first time also show fractioned and/or multiple skeletons and their time go further back than the tenth to the ninth century B.C.E. However, they continued right down to the historical period which confirm that in this area chalcolithic culture transformed into iron age culture. In the North-West Frontier province, iron ores are known to occur in the Panjkora valley; a large deposit of magnetite (Fe_3O_4) has been reported from Gowai in Chitral, where hematite (Fe_2O_3) crystallizing in the rhombohedral system also exists in large quantity between Sanoghar and Mastey; a band of good earthy hematite of 3 to 4 meter, was noted by Middlemin on Sorban Hill in Hazara; while in Wazaristan, local deposits of hematite and concretionary limonite ($2Fe_2O_3.3H_2O$) occur in the Eocene or Mesozoic sedimentary strata. The area continued to be a funnel for movements of people from Central Asia and Afghanistan to the lower radius of the Indus, and to the other eastern regions beyond the Indus is confirmed by the presence of NBPW in this region.

In Baluchistan at Pirak, a site near Mehargarh in Kachhi plain the transition from chalcolithic period to the iron age is clearly indicated by its excavators.[18] A mine of ferrous sulphate i.e. *zagh*, on the eastern slopes of the Nagau range, facing Sanni in Kacchi is present with a large number of Megalithic crain burial sites were discovered from as early as 1876. In Kachhi district the presence of *zagh* in the bed of the Munnali stream about 4.5 kms west of Gajan and near Kumbi west of Kotra. Locally these *zagh* ($FeSO_4$) was used for making a black dye. In Baluchistan no raw iron ores seem to have been procured from outside albit this area is situated along a very important route to Kandahar and further west to Iran in West and north-west India and the Gangetic plain in the east.

In East Rajasthan there seems to be a continuity from the chalcolithic horizon of Black-and-Red Ware into the iron age Painted Grey Ware level. The following NBPW phase probably coincided with the Mauryan and then Indo-Greek. Significant evidence of iron-smelting and manufacture of iron implements is available from a number of sites. Two iron smelting furnaces and forging iron objects were found at Sonari[19] (27°52'; 74°16') in the early phase of the PGW. At Noh[20] (27°16'; 74°58'), crucibles were found in Period III. At Rairh (26°20'; 76°10'), large quantities of slag, which according to excavator, weighed several mounds, and traces of numerous smithies suggested an important iron metallurgical centre. Iron slag was also found on the section cut near the road between Aguncha and Bhilwara.[21] Although large quantities of iron slag were found at Ahar (28°25'; 78°15'), but unfortunately it was not found in the stratified context in south Rajasthan. Iron ore has been found in a number of districts in east of Aravalli i.e. in Mewar region and next in West of Aravalli. In Alwar district occurrence of iron in large quantities have been reported at Rajgarh and Banagarh where the manufacture of iron was in former times a great industry, as is testified by the large hillocks of slags which are to be found in all directions in the area.[22] Iron occurs near Khetri and at Karwar close to Hindaun and 65 kms north-east of Jaipur, around Nimla village.[23] Scattered field of iron ore occur at Lamba in Bhilwara district, while in Udaipur district near Gangrar, about

48 kms north of Chittor, in the hills to south between Bedawal-ka-Pal and Anjani and 61 kms south-east of Udaipur at Nathera-ki-Pal and Thur-Hunder profused quantity of iron ore and the evidence of pre-industrial iron smelting were seen. Since Rajasthan seems to share links, on the one hand, with north-western portion of Indian subcontinent and on the other, with Gangetic plains in the north-east play a pivotal role in the development of iron technology. This can be also confirmed by the presence of largest number of P.G.W. sites in Bharatpur, Jhujhunu, Alwar and in Jaipur districts.

In Indo-Gangetic divide, the workshop of a iron smithy with mud walls and wooden posts at intervals was discovered at Ropar IIIA (38°58'; 76°32'). A furnace was also found with pieces of iron nearby.[24] At Singh Bhagwanpur (39°14'; 75°52') also lumps of iron slag were found.[25] Jauhari and Verma have noted the presence of small deposits of iron ore of an average iron content of 57% in the hilly regions, especially the south of Aravalli. They are found in the Dhanauta and Dhancholi area near Narnaul in Panjab.[26] The region between Indus and Gouja was involved, on the one hand, with the Gangetic plains in the east and, on the other, with the north-west region in the west. Amongst other things, the links with the Indo-Gangetic divide are exemplified by the presence of PGW and NBPW. The PGW horizon is specially very wide in contrast to NBPW horizon.

The archaeological record of Upper Ganga Plain associate iron artifacts with PGW culture and the subsequent NBPW horizon covered all the area with a much greater concentration of sites and thickness of the strata. But iron ore is present in north of it. G.W. Traill[27] noted the presence of iron in various parts of the Kumaun region and the process of extracting iron is extremely simple, a great number of mines are constantly worked. The ore is found near the surface, in extensive strata of rocks, but varying very materially in appearance of different mines. In Garhwal, iron was known to have been mined at Hat, Jainal, Chorpanga and Mokh.[28] In Nainital district of Kumaun region, the chief mines of iron were at Lusgani, Nathuakhan, Gulla and Satbanga, while at Ramgarh, in the same district, there were mainly mines that were worked as late as 1884. The pre-industrial iron industry of this region, from various accounts, also seems to have been well developed. For instance, J.D. Herbert, in his report in 1926 on the mineralogy of the Almora noted that 'for iron nothing is wanting but a proper system of management to render it superior to that of England.'[29] The limestone which is used as flux in iron smelting is also present in the area of north of the Alaknanda, from Lobha patti to Pindar and again to the Alaknanda in Bachhansyum and a range running parallel to the plains and south of the Nayar river. In Nainital, the most important limestone deposits are those in the Kotah Bhabar and in Timilia sot. Limestone is also recorded from the Mandarsu area of Dehradun.[30] It is also to be noted that PGW culture is present in the region of Uttaranchal, thus the source of iron to Upper Ganga plain during PGW phase must be its neighbouring part of Uttaranchal in north and Rajasthan in South. The metallographic study of iron confirm that different segments of upper Ganga plains procured raw materials from different sub-regions. Firstly the Mathura (27°31'; 77°14') and Hastinapur (29°9'; 78°3') relied on the source base of copper and iron available in Bharatpur and its adjacent regions of Rajasthan, while Ahichchhatra (27°41'; 78°16') and Hulas (29°52'; 77°14') among other sites procured copper and iron from the Himalayas to their north. The cultural material seems to be different in various pockets of the Himalaya region. Malari, in the Chamoli district, sixty kms. north-east of Joshimath in the context of a cave burial revealed iron arrowhead.[31] In the Tehri and Nainital district PGW and its associated types have been found at Thapli (Tehri Garhwal), Purola (Uttar Kashi) and Kashipur.[32] Ranihat and Kashipur have also yielded NBPW. In the absence of Neolithic-Chalcolithic horizon in the Uttaranchal, the presence of PGW and a full-fledged iron-bearing level associated with NBPW, undoubtedly suggest strong cultural links with the upper Ganga plains, the epicenter of both PGW and NBPW. The strong possibility of the copper and iron resources of this region being exploited by the Gangetic plain from the OCP—copper hoard horizon ownwards. This association, in fact may have been among the primary factors involved in the cultural developments that took place in the Kumaun-Garhwal Himalaya. The pre-industrial iron smelting tradition of this region also suggests such links. Walton, for instance, noted that the Agarias or Lohars considered their traditional founder to be Kalia Lohar who is said to have forged the weapons of the Pandavas and to whom the Agarias offered five pieces of charcoal before starting work. The context of charcoal confirm the formation of reducing atmosphere in the blast furnace for proper iron smelting.

The archaeological sites of middle Ganga plain are known to be craidle of Buddhist, Jaina and epic material culture in a variety of context. The type-strata of this region is the NBPW which often with iron artifacts. Many megalithic shelters are also found in south of it especially in Mirzapur, Sonbhadra and Chandauli districts in eastern Uuttar Pradesh where iron slag has been found associated with some of them, it is likely that they belong to the same horizon with a long tradition of iron smelters in the yore. Kheradih (25°12'; 83°16'), Ganwaria (27°26'; 83°7'), Taradih (24°16'; 84°16'), Champa (25°10'; 86°17'), Balrajgarh (26°15'; 86°10'), Manjhi (25°16'; 84°17'), Koldihwa (24°54'; 82°2'), Raja Nal Ka Tila (25°11'; 84°10'), Malhar (25°17'; 84°52') have revealed evidence of local iron smelting. Here Raja Nal Ka Tila and Malhar both situated on the Karmanasa river in district Sonbhadra and Chandauli respectively. On the basis of C-14 dates from Raja Nal Ka Tila, the antiquity of iron has been pushed back to 1300 B.C.E. Partly confirmatory evidence of this nature comes from Jhusi near Allahabad where the antiquity of iron goes back to 1100 BCE, but from Malhar, still earlier date to about 1700 B.C.E. if not earlier is reported. At Kheradih[33] in Period II, the excavators in which author was also associated uncovered an iron smith's workshop which in the area around it, was strewn with iron slag. Finished tools of iron and a large sized trough containing a lot of ashy material and some glassy slag were also found as a by product of silca in the workshop's vicinity. At Ganwaria[34], lumps of iron, interpreted as responding local smelting, were noted. Iron slags along with fragments of crucibles were found at Taradih.[35] Champa and Balrajgarh also yielded crucibles.[36] Manjhi[37] yielded iron slag.

Different segments of the middle Ganga Plains were procuring iron ore from different source area. The trace element analysis of Rajghat (25°4'; 83°1') confirm that their source of iron ore was from Singhbhum area of Jharkhand. The south-east Uttar Pradesh also pay a pivotal role in the supply of iron ore to the NBPW people of middle Ganga plain. The Saryupara region relied on the foothills of Himalayas where an iron bearing horizon with NBPW has been unearthed. The Magadh-Anga-Kashi plain looked to the Singhbhum region for iron ore. The importance of the Vindhyan plateau as a source-area of raw materials for the minerally poor Gangetic plains, going back to the mesolithic period, was greatly enhanced during the first phase of the early historic when the urban centers that dotted the plains demanded a large supply of iron. The presence of a good quantity of iron slag at Malhar (in Chandauli district on the bank of river Karmanasa) as also at the nearby site of Lohsanwa is significant and the presence of earthen furnaces within the heaps of iron slag along with terracotta pipe struck with smelted iron provide ample evidence for evolution of iron smelting in the region of meeting point of Gangetic plain and Deccan with local iron ores. This interaction is also evident in the presence of NBPW at sites in the Vindhyas and this region mediated interaction between the Gangetic plain and Deccan.

The second urbanization/NBPW in lower Ganga Plains witnessed the presence of a number of sites in the Bankura, Burdwan, Birbhum, Midnapur and 24-Pargana districts of West Bengal. Additionally, in this phase important centers also come up in the Rajashahi and Bogra areas, many of which have been excavated such as Mahasthangarh (26°12'; 88°16'), Sitakot (25°58'; 89°12') and Banagorh (25°15'; 88°12').[38] Diggings at Pandu Rajar Dhibi (23°35'; 87°39'), Chandraketugarh (22°41'; 88°52'), Mahishdal (23°42'; 87°42') and Bahiri (23°46'; 87°52') have also unearthed NBPW remains.[39] Evidence of local iron smelting comes from Mahishdal where iron ore and slag were found in large quantities[40] and in Period I and II at Bahiri there is evidence of iron smelting. In Period I at Bahiri, an extensive, thick deposit of iron ore and slag was present. In Period II, apart from the slag, a terracotta crucible and piece of a nozzle of a terracotta tuyeres were found. The source of iron ore in lower Ganga Plains is from Santhal Pargana, Birbhum, Bankura and Midnapur districts. In Santhal Pargana abundant iron ores occurring in the trap or in the old sandstone ore even as laterite have been noted by Malley.[42] Iron ores in the Birbhum district have a long history of native extraction and smelting.[43] A lateritic gravel spread has been noted at Muluk while, in the opinion of the excavators of Bahiri, could have supplied the lateritic ore for iron extraction to the site. In Bankura district the iron ore is basically ferruginous laterite, of which there are abundant resources and which were worked by Santhals and Mundas and other aboriginal tribes[44] even in 1877. In Midnapur district also iron ore is said to occur.

In Orissa the high grade hematitic ore bodies of Singhbhum belt extends into Keonjhar and Bonai and Mayurbhanj districts, together with those of Bastar in chhattisgarh are of Upper Dharwar with the remarkable concentration of the metallic iron in them being ascribed to post-Guddapah metasomatic action, to original marine chemical precipitation of the oxides ore of iron by volcanic action and other agencies. These ore-bodies, many of them containing 60-65% of iron, are thought to be the largest and richest deposits of iron perhaps in the world, surpassing in magnitude the Lake Superior ores. They are now estimated to total 21,000 million tones, containing 12000 million tones of metallic iron. Additionally, it is also to be noted that small pisolitic nodules, composed chiefly of hydrated oxide of iron/brown hematite, in the form of laterite, are found all over the states of Orissa; these are also often washed from their matrix, and deposited in sufficient quantities in alluvial formations. The basic affinity of the Orissa plains with the Gangetic plains in the iron age is to borne out by the presence of NBPW at Sonepur in Bolangir district and Sisupalgarh in Puri district. Albeit during Mauryan time, Chunar sandstone was being brought in Orissa from Mirzapur along with raw materials from the Chhotanagpur plateau, but ore of iron was locally utilized.

In Madhya Pradesh, lumps of heavily oxidized iron have been found at Maheshwar (22°11'; 75°36') in his early historic period IV. Also, the rock shelter of Putli Karar in Bhopal have yielded impressive evidence for iron working that includes ores, slag, crucibles along with iron artifacts. The slag chunks were found on the rock shelter floors and almost all the glazed ceramic crucible fragments were discovered in one shelters.[45] Significant ferruginous mineralization has been noted in Raisen, Mandla, Sagar, Jabalpur, Nimar and Bhopal districts which were included in Agaria belt also. In Raisen district, ferruginous laterite is found at Jhamar, where large heaps of slag have also been found. In Mandla district Ball[46] noted native working at the foot of the Amarkantak plateau, near the sources of the Seoni and Bormeyer rivers, in Mowai and near Ramgarh. In Sagar district iron is found abundantly in the form of laterite, of which the most prominent outcrop is found at the summit of Tinsmal.[47] At Hirapur, iron ore occurrences were also testified in the ninteenth century.[48] In nineteenth century C. Gant noted the presence of iron in more than a hundred places, in Jabalpur district of which the principle were Simra, Gogri, Bolia, Agaria, Dalrora, Jauli, Panagar and Lameta. Additionally, some of the localities where a native industry is reported to have flourished are Jauli, where schistone hematites were smelted, and Ghugra east of Sihora, where magniferous iron ore was refined. In Nimar district, hematite (Fe_2O_3) was discovered at Chandgarh where ore was found to be on the surface.[49] The iron age cultural links between the Ganga plain in north and central India are clearly established by the presence of PGW and NBPW occurrence at various sites in Madhya Pradesh. Although the number of sherds, however, is mostly very limited but even at Maheshwar (23°06'; 79°16') on the bank of Narmada in West Nimar district only22 NBPW fragments of which only 13 come from a stratified deposit, were noted. The presence of NBPW/iron in this region seemed to have opened up in different directions with the settlements flourishing in pockets such as Baghelkhand and Sarguja which were largely untouched in the neolithic-chalcolithic horizons is due to local iron ore. The metallographic studies of iron implements from Sooron in Gwalior district confirm very pure form of iron with 0.58% S_2O, 0.45% Al_2O_3, 0.32% CaO, 0.18% MgO. The expertise in the process of cementation during *circa* 600 B.C.E. by which carbon content in the iron could be increased to 0.05% - 0.06%. Plenty of iron slag is also reported from the site.[50]

In Gujarat, early historic iron age archaeological remains are present in Surat, Junagarh, Kheda, Baroda and Bharoch districts in which the important centers are Prabhas Patan (Somnath) (20°53'; 70°24'), Jokha (21°10'; 73°07'), Timbarva (22°2'; 73°14'), Nagara (21°41'; 70°38'), Dhatna (22°12'; 73°52'), Nagal (21°37'; 72°55'), Dwarka (22°12'; 60°0') and Kamrej (22°52'; 73°56'). The iron ores in Gujarat are in the form of laterite which extended from Surat to Bharoch. Limestone strata are also found in the lower part of it which is an important gradient in iron smelting. In Bharoch, iron stone of lateritic origin is distributed over a fairly large area in the Valia mahal.[51] In Surat, Ball[52] noted iron ores in the Balsar and Pardi subdivision accumulation of magnetic sand at the mouths of rivers, especially on Dumas side of Tapti. Additionally Hegde and Mehta[53] ,while exploring for lateritic formations near Dhatva in Surat district, noted two such occurrences within two kms. of the site which contained limonite chunks ($2Fe_2O_3.3H_2O$), like those found in the remnants of ore at Dhatva. In Ambamata area 11th to 12th century A.D. evidence of iron smelting furnaces have been reported. The Gujarat by its geographical position on the eastern coast, from where sea traffic with other countries was articulated, is not effectively reflected in the archaeological record of iron age. No significant urban remains have been noted and even the range of raw materials used is more or less limited to those found within the sub-region itself. Albeit iron is locally smelted by local ores but here in Gujarat iron has nothing to do with urbanisation as it is seen in the Gangetic plain.

In Maharashtra, there is a widespread feature of iron age megalithic burial and habitation site concentrated in Vidarbh region in north and Marathwara region in south. Among the significant excavated megalithic sites in Vidarbha area, are Takalghat and Khapa (20°54'; 78°56'), Mahurjhari (21°14'; 79°30'). Naikund (21°56'; 79°14') and Gangapur (20°16'; 78°15') are noteworthy.At the same time there were a number of early historic sites in the Vidarbh with the merging of the megalithic culture into that phase viz. the significant excavated site in this context are Bhokradan (20°16'; 46°47'), Pauni (20°48'; 79°39'), Songaon (18°00'; 74°35'), Ter (18°20'; 76°10'), Mevasa (19°34'; 74°34'), Paunar (20°47'; 78°41') and Kundinpur (22°55'; 78°05'). The evidence of iron smelting furnace from Naikund[54] is important which were unearthed at a depth of 1.25 meters of 6th century B.C.E. Near the furnace, about 40 kgs of iron slag was found along with the Micaceous Red Ware similar to those recovered in the habitational deposits and megalithic stone circle. Two clay made tuyeres which attached with the furnace which help in blasting of it during smelting process were also recovered. Terracotta crucibles were also found at Bhokradan[55] and Nevasa[56]. In Maharashtra, the chief sources of iron are laterite and magnetic-sands of rivers draining the trap districts both of which are largely drawn up by the inherent blacksmiths. Good quality of iron ore is apparently found near Mansar in Nagpur.[57] The Naikund in the same district also evidence the presence of iron ore in a *nala* about 1 km south-east of the excavated site. Here a magniferous belt of iron, which is 3 kms away from the Naikund excavated site is seen. This is significant in the light of the iron analysis of Naikund and are showed crystalline micaceous hematite quartzite, a type generally associated with mangnese ores confirm that at the time of Megalithic period, the populous exploits local iron ores. The Chandrapur, Bhandara, Ratnagiri and Amravati districts of Maharashtra have richest deposit of lateritic iron ore. Due to this very fact, the maximum number of iron age settlement is seen only in those districts of Maharashtra where iron ores are frequently reported. Thus the Early historic Maharashtra seems to have been relatively self-sufficient as far as iron ores were concerned.

The cultural horizon of the North Karnataka in the post-Neolithic Chalcolithic phase was dominated by the appearance of iron age megaliths viz. Brahmagiri (Dt. Chitradurga), Mastki (15°57'; 76°38') and Piklihal (16°7'; 76°34') in Dt. Raichur, and Hallur (14°20; 75°37') in Dt. Dharwar. Various smelting areas were found around Piklihal which, according to Allchin[58], may possibly go back to the peiod when iron artifacts were first manufactured at the site.

Like the north, south Karnataka have a number of megalithic settlements found primarily in Mysore, Kolar and Chikmaglur districts. The representative site is T. Narsipur in Dt. Mysore which exploited magnetite iron ore of the same district.[59] In Chikmaglur district important iron ore deposits are found at Attigundi, Galikhere, Gundikhan, Rudragiri and Jensarigudda. In Kolar, iron ore is found in the form of laterite in disconnected patches on flat-topped hills around Sidlaghatta, Srinivaspur, Malur and Jangamkote.[60] Like north Karnataka, in south also no iron ores seem to have been procured from outside it. The strip of land bound by the Western Ghats on the east and the Arabian sea on the West i.e. Kerala, where megaliths are much later than those in Karnataka. This is based on the megalithic excavation of Porkalam (10°40'; 76°4') in Dt. Trichur of Kerala by Thapar.[61] Here in the state of Kerala iron ore occurs in the districts of Calicut Kozhikode, Kottayam, Malappuram, Palaghat, Quilon, Trichur, Cannanore and Coorg. Laterite is the primary iron ore of this area and this term was first used for material from Malabar by Francis Buchanan in 1800. In his own word "The ore is found forming beds, veins or detached masses in the stratum of indurated clay that is to be afterwards described, and of which the great part of the hills of Malabar consist. This ore is composed of clay, quartz in the form of sand, and of the common black iron sand. This mixture forms small, angular nodules closely compacted together and very friable. It is dug out with a pick-axe."[62] Thus in Kerala, the absence of neolithic-chalcolithic horizon, the full-fledged iron age horizon in the last centuries of B.C.E. there was probably an outcome of influences emanating from Karnataka. In Andhra Pradesh iron age megalithic remains being found in the Guntur, Krishna, Mahbubnagar, Prakasam, Nalgonda, Srikakulam, Bhadrachalam, Kurnool, Cuddaph, Rangareddy, Karimnagar, Warngal, Hyderabad and Adilabad districts. In the area of Andhra Pradesh crystalline and schistose rocks of the Cuddaph systems enclose at some places ferruginous deposits of an extraordinary magnitude. Here iron ore in the form of Magnetite-quartzite is present in Adilabad, Kurnool and Prakasam districts, while hematite in seven square kms occurs in Krishna district. In

Cuddaph district low grade iron ore is reported.[63] Unlike Karnataka, Kerala and Tamil Nadu, Andhra Pradesh is exemplified by the presence of NBPW in the late phase of iron age. At the same time major early historic iron age strata of the peninsular south are found in Andhra Pradesh.

Tamil Nadu have no cultural gap in-between the neolithic horizon merged into the iron age which, as in the rest of south, was dominated by the megalithic class of monuments whose earlier chronological limit seems to be pre-Sangam epoch. Megalithic sites are found in South Arcot, North Arcot, Tanjavur, Chingleput, Salem, Tirunelvelli, Coimbatore and Madurai districts. Here, the early historic period which is a continuation of the iron age, marked by the presence of russet-coloured ware seems to have commenced during the 4th-3rd century B.C.E. In Tamil Nadu, iron ore deposits in the form of magnetite-quartzite bands have been traced in Salem, Dharmapuri, South Arcot and Tiruchirapalli districts.[64] Iron slag has been found near Appukallu[65] and at Odugattur.[66] At Guntur, the evidence of local iron smelting includes several pieces of terracotta objects which may have served for blasting the furnaces for iron smelting is reported.[67] Thus, there were a number of flourishing iron age sites in Tamil Nadu with rich artifacts, which in most cases, could avail themselves of the iron ores found within the same region.

Thus, if iron ore is much more abundant (5% of the earth's crust) than copper (0.5% of the earth's crust) and its ore like hematite were used in upper Palaeolithic cave arts as colouring agent throughout world in the later phase of Plaeistocene to early phase of Holocene, then why its uses comes later in the history of Pyrotechnology? The reason is only due to its inherent technical complications of extraction from the ore. The density of copper is 8.89 gm/ml and melting point is 1083°C. Thus extraction of iron from its ores need at least 500^{0}C more temperature than the temperature needed for the extraction of copper from its ore. The pottery baking in their kiln can be started from 800°C to 1200°C, which were known to the populous of Neolithic time. Although the pure Kaolinite [$Al_2Si_2O_5(OH)_4$] of the clay by which pottery was made melts at 1770°C, far beyond the reach of early kilns. But the inclusion of impurities in the clay which act as fluxes, the clay particles will sinter together at much lower temperature—800°C to 1200°C for most ancient pot clays. It is however unlikely that even this temperature was reached in the majority of early kilns. The usual temperature of the potter's kiln was from 500°C to 800°C during Neolithic period. But kiln temperatures improved in the Bronze Age culture throughout the ancient world. Thus the increase in temperature of the potter's kiln make an important factor in the innovation and evolution of copper technology. The copper is either extracted from its sulphide ores or by oxide ores, but the smelting of oxide ores can be easily in contrast to its sulphide ores, because sulphur is a reactive non-metallic element and melted at 444.4°C, while copper melted at 1083°C. Thus removal of sulphur during extraction process of copper is technically difficult. On the other hand, the oxide ores of copper involve the least complicated process. Albeit Harappan utilizes sulphide ($CuFeS_2$) ores from Khetri mines. The chalcopyrite ($CuFeS_2$) have both copper and iron, but during extraction of copper from this sulphide ore, iron was not extracted in the yore by which the impurities of iron in the sulphide ore, the copper-bronze artifacts have iron in their composition as it was seen in the chemical analysis of various copper artifacts of Indian chalcoliths. Both copper and iron are transition metal having two oxidation number.[68] Iron pyrite (FeS_2), though plentiful, is not an economic source of iron due to difficulty in reducing the sulphur content to the level required ($<0.05\%$). In the case of copper it has +1 and +2 oxidation number but iron has +2 and +3 by which it is more reactive to form ore body and the dead rock or gangue.[69] The ore body is the name given to the natural compounds of iron, most frequently its oxide in the form of hematite (Fe_2O_3) and magnetite (Fe_3O_4), found in the ore. Practically, all the iron ore is reduced from its chemical compounds in the blast furnace[70] and goes over into the pig iron.[71] The gangue of iron ore may be of various chemical composition. Most frequently it consists of quartzite (SiO_2) or sandstone with an admixture of clayey matter ($Al_2O_3.2SiO_2.2H_2O$) and less often of limestone ($CaCO_3$) or dolamite ($CaCO_3.MgCO_3$). In blast furnace the gangue is converted into slag, thereby separating the pig iron. Here slag is the mixture of flux, impurities which floats on the surface of molten metal. Its purpose is to act as repository for impurities or it may also be adjusted to effect the removal of impurities and to protect the metal from contamination by the atmosphere. In the case of iron a certain amount of impurities are always present in the form of arsenic, sulphur and phosphorous. In the modern time the best iron ore is which contains a negligible amount of harmful impurities and the minimum amount of gangue in a given condition. The gangue in hematite is most often quartzite while in magnetite it is of a siliceous nature with an admixture of a certain amount of other oxides. It is more difficult to reduce magnetite than other iron ores; in some cases it contains much sulphur (upto 1.5% to 2%) and sometimes such objectionable impurities as zinc which destroys the furnace linings. The gangue of siderite ($FeCO_3$) ore consists of SiO_2, Al_2O_3 and a small amount of MgO. Certain siderite ores contain a sandy-argillaceous gangue. Limonite ($2Fe_2O_3.3H_2O$) is not a firm ore of iron because it contains phosphorous, while siderite possesses the highest reducibility of all iron ores. Before being charged into blast furnace, siderite ore is usually calcinated according to the reaction as—

$$FeCO_3 \rightarrow FeO + CO_2$$

This removes the hygroscopic water and the carbon dioxide (CO_2) combined with the iron. The siderite ore become very porous and brittle; it is easily crushed with forming much dust, as it is seen in the case of limonite. In the iron ore, sulphur, phosphorus and arsenic is harmful, but at the same time it contains other useful metals besides iron which pass into pig iron in the blast furnace process of iron smelting. These metal alloys the pig iron and improve many of its properties. Among the more valuable composite iron ores are chrome-nickel iron ore and chromium iron ore from various deposits.

The oxide of ores of tin, lead and copper could be easily reduced and melted, because in comparison to iron they have low melting point. Thermodynamically iron oxide can be reduced to iron at any temperature above 800^{0}C, but since the melting point of the metal is 1540^{0}C, the chalcolithic furnaces having a temperature in the range of 1100-1200^{0}C could result only in solid state reduction of iron oxide, while proper smelting need solid solution reduction in which the microstructure of the molten metal consists of similar crystals with no evidence of the separate existence of either alloying element or of intermetallic compounds. Thus solid state iron in smelting process caught up in the reduced spongy solid mass used to be molten slag which had to be squeezed out by hammering and forging. Here forging means plastic deformation of a metal by squeezing or hammering, to produce a required shape. The ancient smith took a long time to understand the intricacies of solid-state reduction and the fact that the soft mass of iron could be hardened and made useful only through prolonged contact with charcoal which is known as process of carburization in which increasing the carbon content of a metal by heating it in contact with a carburising medium (which may be solid, liquid or gas) usually for the purpose of producing a hardened surface by subsequent quenching. The carbon steels are the result of carburization. Forming cementite (Fe_3C) in the steel, it both strengthens it and renders it amenable to heat treatment. The cast iron is also a result of carburization which is an iron-carbon alloys (2-4%C), all cast to shape, but some treated to provide ductility, while others may show high hardness, corrosion resistance etc. In contrast to cast iron, wrought iron have no carbon but containing numerous iron at silicate inclusions which is noted for its toughness, weldability (by forge welding methods) and corrosion resistance. It is made by melting pig-iron in the furnace and refining it by the addition of iron oxide to oxidise the impurities into a slag. As the purity of the iron increases, so does its melting point, so that the iron becomes a party mass containing a considerable quantity of entrapped slag. The advantages of wrought iron is due to its softness, easily shaped and welded. In Indian context Mehrauli iron pillar (6th century A.D.) of Delhi, Dhar iron pillar of Malwa (12th century A.D.) and iron beam of Lingraj temple in Bhuvaneshwar and Sun temple of Konark of 11-12th century A.D. were made by wrought iron. Since wrought iron is not too hard, therefore, it was not used in making tools or weapons which demanded hard cutting edges. The ethnometallurgical studies of iron technology in Chhotanagpur and Bastar[72] confirm that by the process of forging of pig iron, wrought iron were obtained in ancient time. Thus iron was probably first produced, in the form of wrought iron, as an accidental byproduct of lead and copper smelting, because in the ancient Near East the iron ore hematite was often added to siliceous ores of lead and copper as flux or by the formation of ash mound in South Indian Neolithic on laterite base. The property that sets iron apart is its response to heat treatment. Like any other metal, iron can be hardened to some extent by cold working; unlike other common metals, it can be hardened dramatically by quenching. Quenching is a process of rapid cooling in water from temperature above red heat, 725°C the exact compromise between hardness (which entails brittleness) and toughness best suited to any particular application can then be achieved by tempering which reducing the hardness by reheating at a temperature between about 200 and 400°C. It should be noted that the presence of carbon is essential in both tempering and quenching, because quenching has little effect on pure iron. Before the mastery of carburization and quenching, iron could not be made to equal the performance of a work-hardened bronze, and its use is likely to reflect only the abundance of its ores and consequent cheapness of the metal. Carburized and quench-hardened, however, iron becomes immensely superior to bronze. It might be added that the techniques required to make good steel cannot borrowed ready made from the bronze workshop: bronze and copper are hammered cold rather than hot, because red-hot bronze may even shatter when struck and they do not form alloys with carbon, and quenching has no effect on them (heating copper or bronze) annals it i.e. soften it, regardless of whether the subsequent cooling is fast or slow.

Benefication of the iron ores before smelting involves both enrichment of the metal content and the improvement of the physical characteristic through crushing, squeezing and washing. Over 200 studies in the National Metallurgical Laboratory, Jamshedpur have indicated that washing of the iron ores before smelting is an absolute necessity to solve the problem of sticky and wet ores and to yield sized lumpy ore for efficient blast furnace operation. It is pertinent to point out that during ethenic smelting of iron, the ore were thoroughly washed by water which confirm its ancient practice too in Indian context. Washing of the iron ore, if employed optionally, it can decrease their silica and alumina content. Minerological studies have indicated that alumina, control of which is essential for obtaining a fluid slag in the blast furnace, may be present in the Indian ores in several forms. It may exist as fine clay and adherent material or as lumpy lateritic material. While the clay materials can be washed, the lateritic material may pose greater problems as it may be interspersed in the ore body.

At Dhatwa, in Surat district of Gujarat the evidence of benefication of the iron ore is observed by Hegde[73] during Early historic time. Here the analytical data indicate that the deep red iron ore are hematite while the light red ore are limonite. It appears likely, therefore, that the limonite collected from the surface laterite formation was roasted at the site to convert it into hematite before smelting. This indication of roasting of the ore at Dhatwa is not without significance. Limonite ($2Fe_2O_3.3H_2O$) has a variable colour, ranging from yellow to light red depending upon its hydration. When it is roasted, as the data for loss on ignition indicate, much of the water content is expelled together with CO_2 and other volatile components of the ore. Washing and roasting of the ore therefore increases in iron oxide content weight per weight and renders it more porous, fragile and hence easier to crush more suitable for reduction. The same practice with laterite is adopted by ethenic people of the Bastar in Chhattisgarh. In contrast to Chhattisgarh, in Gujarat laterite ore was not used for iron smelting in the yore, because no mining was observed in basal laterite formation. On the other hand, the pieces of limonite in the surface of laterite formation were found, lying scattered around and could be easily collected.

Smelting of iron from its ores need a very high temperature, thus fuel play a pivotal role in the development of pyrometallurgy. All fuels are of organic nature; their combustible constituents are carbon, hydrogen and various compounds of these elements (hydrocarbons). Fuels also contain certain amount of oxygen and sulphur, in which sulphur is the most harmful constituent of fuel since in the process of smelting some of it may dissolve in the metal thus impairing its properties. Fuels used in modern metallurgy, should contain the least possible amount of inert mineral admixtures (ash) which are incombustible and reduce the calorific value. Only solid fuels are employed in iron smelting which includes charcoal, coke and coal. India has 29 billion tonnes of coking coal reserve of which 18 billion tonnes is miniable, but coal was invented after Industrial Revolution, thus it was not used as fuel in pyrometallurgy before 18th century. Thus in Ancient and Medieval time charcoal was used as a fuel and not coal. The charcoal was obtained by the dry distillation of wood (heating without the admission of air) in stalls. It contains 80% to 90% solid (non-volatile) carbon, 10% to 18% volatile matter including moisture, 0.6% to 1.2% ash and a negligible amount of sulphur. The carorific value of charcoal ranges from 6500 to 7500 kcal per kg. The same is the calorific value of coke too i.e. 6500 to 7500 kcal per kg which is heigher than the calorific value of coal (4500 to 8000 kcal per kg.) Charcoal is highly porous (upto 70% it burns well) but has a low comprehensive strength (20 to 40 kg per cm). In modern time too charcoal is used in small-volume blast furnaces for making high grades of cast iron with a low sulphur content. Fuel plays an important role in the blast furnace process. Its function is twofold; first, to evolve enough heat in its combustion to attain the high temperature required for reducing the ore and for melting and superheating the pig iron and slag that are formed; and second, to supply carbon that directly participate in the ore reducing reaction because every smelting[75] is a process of reduction and not oxidation. Before invention of coal, wood-charcoal was thus probably used as a fuel for smelting. By itself, however, wood is unsuitable for smelting as it contain too much oxygen and hence does not make a good reducing fuel. Unfortunately in Indian archaeology the proper analysis of charcoal that obtained from excavated furnaces were not studied by which it is not possible to say what wood was used to produce the charcoal needed.

Flux is the next important material in pyrometallurgy of ferrous whose functions are to lower the smelting point of the ore gangue and to fuse it with the ash of the fuel. The most commonly used blast-furnace flux is limestone ($CaCO_3$); dolomite ($CaCO_3.MgCO_2$) is employed less frequently in iron smelting. In some case siliceous alumina fluxes are charged into the blast furnace e.g. in the region of Chhattisgarh and in others, only silica sand was used as flux. The quality of a definite type of flux required for operation determined by computations based upon the chemical composition of the gangue and ash, and upon the requirements made to the physiochemical properties of the slag formed in the blast furnace. Theoretically blast-furnace fluxes should contain a minimum amount of the harmful impurities—sulphur and phosphorous. Silica is an undesirable constituent of limestone because it decreases the fluxing power and increases the volume of slag in the blast furnace. At present limestone is an important flux material for the iron and steel industry of India. Current estimates put the total reserves of limestone at about 69 billion tonnes of which only 20% can be exploited for use in the iron and steel industry. While limestone for iron making changed into blast furnace and can have a certain flexibility with respect to the desirable level of silica. Each additional 1% silica reduces the available lime by 2.5%, increase slag volumes and forms a thick insulating layer of very low levels of silica are thus desirable in limestone used as flux in iron smelting. But in ancient time in India, lime as flux was rarely used, the slag was often tayalite ($2FeO.SiO_2$) whose melting point is in the range of 1150-1200°C. The slag coating the metal further retarded carbon pick-up. In the ancient process considerable amount of iron oxide remained in the slag unreduced, iron recovery was low but the purity of the extracted metal was high. The high oxygen activity of the tayalite slag in ancient time acted as a control parameter in lowering the carbon content of the metal. In contrast to it in the modern blast furnace, productivity and temperature being high and limestone being used as flux, tayalite concentration in the calcium silicate or calcium alumino-silicate goes down, oxygen potential decreases and hence the carbon pick-up in the metal is increased. Thus the indigenous technology had low productivity and thermal efficiency and yet produced pure malleable metal which was used to forge a large number of locally useful materials.

In archaeometallurgical context slag which usually find out during archaeological excavation deposit throw light on the process of smelting of the metals in the past. Slag is the mixture of flux, impurities (and sometimes also refractories) which floats on the surface of a molten metal due to its less density than that of pure metal. Its purpose is to act as a repository for impurities and these impurities predict past history of its technological formation. It is a by-product of the refining of metals from its ores, most importantly in the reduction of iron in a blast furnace. It was formed because the iron ore has impurities such as silicon dioxide, which can react with the calcium oxide produced by thermal decomposition of the limestone added to the furnace. This high temperature reaction produces the slag calcium silicate:

$$CaO(\text{flux}) + SiO_2\ (\text{impurity}) \rightarrow CaSiO_3\ (\text{slag})$$

The molten slag is less dense than the molten iron and floats on the surface from where it can be tapped off. If the flux is not added during melting process of iron then silcon dioxide as impurity can't be removed from the iron because the melting point of SiO_2 is 1730°C, while that of iron is 1540°C. Flux help in lowering the melting point of SiO_2 upto 800°C, by which it removed from the molten metal.

Pyrochemistry of the blast furnace started with the optimum ratio of charcoal, ore and fluxes, by which proper smelting can be done. Theoretically in modern smelting furnace of iron the space is divided between coke, ore and fluxes in the ratio of 10:3.6:1. But how this ratio was obtained in the yore? Ancient literature of the India and outside India is totally mum on this issue. The ethenic study of the iron smelting done by the author[76] in the region of Chhotanagpur and Bastar confirm that the ratio between ore and charcoal was in 3:1, which was very near to modern ratio of 10:3.6.

The pyrochemistry of the whole smelting operation of the iron ore could be divided into three main stages:

(1) The combustion of charcoal to CO-CO_2 gases and the solid state reduction of oxide ore of iron to metallic iron in the furnace. This solid state reduction of iron started at 800°C, if the ratio between CO:CO_2 is 4:1 if the ratio between two reducing gases i.e. CO and CO_2 is changed from 4:1 in the furnace then the reduced iron is again reoxidised into its ore howsoever temperature of the furnace may be increased to whatever degree. This is the most complicated stage of blasting. At the top of the furnace throat the temperature of the rising gases equals 300°C to 350°C and it gradually increases in the lower level of the furnace upto 1600°C by the combustion of charcoal which burns according to the reaction

$$C + O_2 \rightarrow CO_2 + \Delta Q.$$

Since it is an exothermal reaction, thus a large quantity of heat is evolved in this reaction and carbon dioxide (CO_2) is formed which is almost completely decomposed upon carbon monoxide (CO) with contact by hot carbon:

$$CO_2 + C \rightarrow 2CO$$

Thus, the gaseous phase formed in the lower part of the furnace acquires intensive reducing capacity for solid state reduction of iron oxide ore upon contact with the rising hot gases the charged materials (charcoal + ore + flux) first loose their hygroscopic and then their chemically combined water. Due to their moisture loss (dehydration) the lumps of the charge decreases in weight, become more porous and sometime crake. The combined water (hydrated) is driven off beginning with 102° – 105°C and in some cases this process is completed only at 450° - 500° C (e.g. from lump of clayey matter—$Al_2O_3.2SiO_2.2H_2O$). The hydrated iron oxide ($nFe_2O_3.mH_2O$) in hematite ore is converted into ferric oxide (Fe_2O_3) when ore is heated to 200° – 500°C. Dehydration of the charged materials begins in the furnace throat and is usually finished in the upper half of the stack. Here, the remaining volatile matter is removed from the charcoal (as H_2, CH_4 and other gases).

(2) Thermal decomposition (dissociation) of the carbonates in the flux and in certain kinds of iron ores (siderite) occurs in the middle and lower portion of the furnace. The temperatures at the beginning and completion of the decomposition depend upon the nature of the carbonates and the size of the lumps; e.g. the decomposition of the limestone to form lime, according to the reaction

$$CaCO_3 \rightarrow CaO + CO_2$$

Take place at 900°-1000°C, but siderite decomposes to form magnetite by the reaction

$$2FeCO_3 \rightarrow Fe_2O + 2CO_2 + CO$$

at low temperatures (400° – 500°C).

The carbondioxide evolved in the decomposition of $CaCO_3$, $FeCO_3$ and other carbonates dilutes the blast furnace gas and decreasing its carbon monoxide concentration. When their moisture has been driven off, the lumps of iron ore undergo reduction which leads to the metallic formation of iron.

The direct reduction of FeO to Fe and formation of FeO rich tayalite ($2FeO.SiO_2$) slag in tyuere zone. The slag metal reaction controlling the carbon and metalloid content of the iron sponge. The reduction of Fe_2O_3 to Fe being topo-chemical which get reduced into three stages in the order $Fe_2O_3 \rightarrow FeO \rightarrow Fe$

$$3Fe_2O_3 + CO \rightarrow 2Fe_3O_4 + CO_2;$$

$$Fe_3O_4 + CO \rightarrow 2FeO + CO_2;$$

$$FeO + CO \rightarrow Fe + CO_2$$

The whole formation of Fe from Fe_2O_3 is irrespective of the chemical reaction being carried out by solid carbon (direct reduction reaction) or by CO gas (indirect reduction reaction).

(3) The optimum thermo-chemical condition required for the reduction process could be determined by iron-oxygen-carbon equilibrium diagram[77] plotted as a function of CO/CO_2 ratio temperature. From this equilirbium diagram it is obvious that as temperature increases, the stability of CO gas also increases with respect to CO_2 and beyond 1000°C only CO gas is found in the gas generated at tuyere level. Hence, the rate controlling step for the combustion of carbon is $CO_2 + C \rightarrow 2CO$

The CO/CO_2 equilibrium line crosses the FeO-Fe phase boundary at point 'P' which represents the minimum condition for the reduction of iron oxide to iron. Based on the studies on reduction kinetics of iron oxide in CO/CO_2 gas mixture the optimum temperature required for fast reduction is between 900 to 1100°C and in this temperature range the $CO:CO_2$ is 80:20 i.e. 4:1.

Thus the equilibrium reduction reaction could be written as:

$$Fe_2O_3 + xCO \rightarrow 2Fe + 3CO_2 + (x\text{-}3)CO$$

where the ratio between $CO:CO_2$ must be 4:1. Thus the reaction feasible under equilibrium condition is

$$Fe_2O_3 + 15CO \rightarrow 2Fe + 3CO_2 + 12CO$$

As the iron ore descends down to the shaft, it gets partially reduced to FeO and Fe by the chemical reaction with hot carbon mono-oxide rich gas as well as by the preheated charcoal. In the bottom part of the furnace at 1100°C to 1300°C the following chemical reactions take place:

$$2FeO + C \rightarrow 2Fe + CO_2;$$

$$CO_2 + C \rightarrow 2CO$$

The total reaction being

$$FeO + C \rightarrow Fe + CO$$

At the same time FeO begins to react with SiO_2 and other gangue material present in the ore to form FeO-rich tayalite slag. $FeO.SiO_2$ phase diagram confirm that FeO forms an eutectic phase with tayalite at 1175°C. Hence, the slag will be fluid enough even at 1300^{0}C to drip through the solids and react with charcoal ash present in the tuyere zone before getting collected at the bottom of the furnace. As the iron moves further downwards, it picks up carbon to form cementite (Fe_3C) i.e. carburization of sponge iron take place by the reaction.

$$3Fe + 2CO \rightarrow Fe_3C + CO_2$$

Thus theoretically metallic iron formed in the blast furnace is initially in the solid state (spongy iron) since it has a high melting point (1540°C), but in the presence of carbon monoxide when cementite[78] is formed its melting point is lowered to 1150°C – 1200°C. The result is the carburized iron (containing upto 1.8 – 2% C) melts and runs down. In the West Asia and Europe, where cast iron does not seem to have been made deliberately before the 14th century A.D., the manufacture of steel prior to that time began with wrought iron and added carbon to it by a simple process of cementation in which a piece of wrought iron deeply embedded in a charcoal fire is converted to steel by prolonged heating which allows it to absorb small amount of carbon by solid-state diffusion. Although smiths stumbled on the cementation process before the end of the 2nd millennium B.C.E., the essential alloying role of carbon was not realized until the 18th century A.D.: until then 'steel' was simply the name given to a mysteriously fine iron that a good smith knew how to produce. The ancient smith certainly never guessed that his charcoal fire was adding carbon to the iron to form an alloy; indeed, he was more likely to believe that he was purifying the iron in 'the refiner's fire'. Thus the success of the ancient pyrotechnology of iron depended very much on the selection of the raw material and the maintenance of the critical balance between the reducing and oxidizing reaction in the optimum zone. Based on these considerations, the importance of the control of air blowing rate is the most important factor in overall process control and based on the calculations of the material and heat balances. For this, the air blowing rate was modified based on the desirable colour (blue) and height of the carbon monoxide gas flame above the furnace width and also the observation of the slag properties. If the air blowing rate was increased, then it invariably resulted in the increase of the furnace temperature and production of molten while cast iron which came out along with slag. It was considered to be a bad event regarded as a prophetic sign, because in ancient times this highly brittle material could not be shaped by forging. It was a waste material till the processes of cast iron casting malleablization were developed in China and the process of wootz steel production using white cast iron was perfected in South India.

Alloy is solid solution combination of two or more elements of which at least one is metal. Most pure metals can be strengthened by small additions of other metals or non-metals which improve the physical properties of the resulting metal. Alloys are not simple mixture but complex crystalline structures which may differ considerably from any of their constituents can bring about significant changes in the properties of the material. Both bronze (alloy of copper and tin) and brass (alloy of copper and zinc) are considerably harder than copper. The most striking example is steel, which is essentially an alloy of metal (iron) and non-metal (carbon). Pure iron has a tensile strength[79] of the order of 14 tons/sq. inch and a hardness of approximately 70 Hd/30. The addition of less than 1% of carbon has two important effects. It both increases the strength and hardness of the material in the slow-cooled (i.e. softened) condition, and permits it to be hardened by quenching from a very high temperature. It therefore permits the production of a material which can be shaped and machined in the softened condition, and then hardened by heat-treatment. With 0.5% carbon, the tensile strength in the softened condition is in excess of 30 tons/sq. inch with a hardness of 160Hd/30. When fully hardened, this material can attain a hardness in excess of 650 Hd/30. A corresponding figure for tensile strength at this hardness is of limited value since the material is so brittle that fracture in tension is highly dependent upon axiality of loading (to avoid bending) and surface finish. The compressive strength is, however, very high.

Pure iron melts at 1536°C, a temperature inaccessible before the 19th century A.D. Hence in antiquity iron could be reduced from its ore in molten form only if during the smelting process it absorbed non metal carbon from a charcoal fire, thereby forming an alloy with a lower melting point. When the carbon content reaches 3 to 4.5% the melting point falls to about 1150^{0}C, not much higher than the melting point of copper (1083^{0}C) or the temperature of a good pottery kiln. Generally the iron produced from the bloomery furnace had a maximum carbon content of 0.4% which had poor hardenability, but definitely much harder than pure iron. Thus invention of the carbon-iron alloy was not intentionally but it was due to unable to making high temperature of the furnace by ancient smithy. The other range of iron-carbon alloy produced in the ancient India contained 0.7 to 2% carbon. Ghosh[80] have reported an iron sickle from Barudih, in Singhbhum district of Jharkhand dated back to 810 B.C.E. which contained 0.35% carbon. The microstructure contained of ferrite[81] and pearlite[82] and it also showed the presence of tempered martenstite and the sharp edge. The analysis of iron objects from Taxila in Pakistan by Hadfield[83] confirm the presence of 1.3% to 1.5% carbon during 300 B.C.E. According to him many features of these iron objects suggested deliberate production of alloyed steel. Thus the iron object analysis data from archaeological excavations confirm that the Indian artisan moved from the impure bloom (*circa* 1800 B.C.E.) to wrought iron (*circa* 1000 B.C.E.) and then on to 'steeled iron' (*circa* 400 B.C.E.) over a period of 1200 years. The process of steel manufacturing by carburization of wrought iron in ancient India could be carried out by any one of the following process:

(a) carburization of wrought iron inside the bloomery furnace by charcoal.

(b) Carburization of wrought iron during the secondary refining i.e. quenching
(c) Carburization by solid-state diffusion in which packing of the iron object in charcoal powder and heating to a high temperature for a long time. (Wootz process)
(d) Cementation of iron having different carbon content.

By all the above processes the carbon content in the iron could be raised upto 0.9% which could be easily hardened, but in these processes the slag inclusion present in the bloomery iron could not be removed and they caused lower tensile strength as well as fracture toughness. By *circa* 400 B.C.E. a better quality of steel produced in Andhra Pradesh as 'seric' and later known as 'Wootz' steel, a name derived from the Telugu word '*Ukku*'. This steel containing between 1% to 2% carbon had no parallel in the world and its production and processing had reached its zenith during 18th century A.D. By 17th century A.D. large shipments running into thousands of wootz steel ingots were exported from Coramandal coast to the West Asia and Europe. Although the processes of wootz steel making were developed and perfected at Konasamudram near Hyderabad, at Salem in Andhra Pradesh, Malabar in Kerala and Golkunda in Karnataka, by the end of 18th centry A.D. wootz steel objects were being made at many centers of north India like Amritsar, Lahore, Agra, Jaipur, Gwalior and many other places but none of these centers are surviving today and 2500 years back ancient technology has been forgotten in India. Technically the production of 'wootz' would require a very high quality of iron ore which explains the exceptionally long and localized monopoly of the smelters of Andhra Pradesh and Karnataka in its production. The slow cooling of the bloom metal after absorption of carbon from the charcoal would lead to a hyper-eutection ferro-carbon alloy[84] with partially and heterogeneously spheroidised cementites having a carbon content normally running between 1.2% to 1.8%. To this day, attempts are being made to mass-produce this type of ferro-carbon alloy because of its combination of flexibility and hardness which is not available in modernized steel. The Russians[85] tried to copy it in the 19th century and it was the subject of numerous studies by Europeans in the 19th and 20th centuries. Attempts by Anosov to study the surface manifestations of these structures microscopically also lead to the birth of metal crystallography.[86]

Notes and References

1. Precambrian is that period of time during which the earth's crust was formed and first life appeared. The terms Archaeozoic, Archaean, Azonic, Cryptozonic, and Proterozoic have been used as a synonyms for this period. The time duration of the period is probably not less than 4000 million years and covers 90% of the gological time scale of the earth.
2. The gravitational capacity of a substance per unit mass is known as specific gravity.
3. The minerals of little economic value which are associated with metallic ores.
4. In *Rigveda*, 32 references of *ayas* in the form of *shyam* and *lohit* is seen, but the term *loha* was of later Vedic tradition e.g. in *Vajsanai Samhita* (XVIII.13); *Satapath Brahmana* (V.4.12); *Chandogya Upanishad* (VI.1.6).
5. Metasomatism is a metamorphic process whereby existing minerals are transformed totally or partially into new minerals by the replacement of their chemical constituents. This occurs by the introduction of capillary solutions, originating either outside or within the rock, that are reactive with the existing minerals.
6. Ore body is a continuous, well-defined mass of material or sufficient ore content to make ore extraction economically feasible.
7. It is a residual deposits which are formed due to metasomatic surface replacement of Dharwar slates and schists by manganese/iron bearing solutions. It consists of hydrated aluminium iron oxide. They are irregular in distribution, occurring as caps on the outcrops of the Dharwar rocks, as is evident from the peculiar nature of their origin. It is a porous, pitted, clay-like rock with red, yellow, brown, grey and mottled colours, depending in some measure on the composition. It has a hard protective limonitic crust on the exposed surface which is generally irregular and rough. When dug up, the fresh material is comparatively soft and easily cut by a spade or a saw. When the fresy rock is exposed to air, it is quickly dehydrated and becomes quite hard like brick stone. When it has been exposed to air and has hardened into rock, its colour becomes black. In accordance with the relative

amounts of the elements present, laterite may be called ferruginous, aluminous and magniferous. In general, ferruginous laterite is red to red-brown in colour, the aluminous ore grey or cream and magniferous ore dark brown to black.

8. It is a compound of Titanite ($CaTiSiO_5$) and iron, which is widely distributed as accessory mineral in acid and intermediate igneous rocks, in their associated pegmatites and several metamorphic rocks.

9. It is a compound of vanadinite i.e. $2[Pb_5(VO_4)_3Cl]$ and iron occurring as globular masses encrusting other minerals in lead mines.

10. Lawrence, W.A. (London, 1895). *The Valley of Kashmir*, p. 62.

11. Chakraborti, D.K. (Delhi, 1984). "The Archaeology of Iron Age and Early Historical Period of Kashmir." *Man and Enviroment*, Vol. VIII, p. 112.

12. Kamili, M.H. (Sri Nagar 1967). *Census of India*, 1961, Vol. VI – Part IX, Census Atlas of Jammu and Kashmir.

13. Sharif, M. (Karachi, 1969). "Excavation at Bhir Mound, Taxila. *Pakistan Archaeology*, No. 6, p. 12.

14. Coggin Brown, J. and Dey, A.K. (London, 1955). *India's Mineral Wealth*, p. 190.

15. Stacul, G. (Rome, 1966). "Preliminary report on the pre-Buddhist necropolises in Swat (W. Pakistan)." *East and West*, No. 16, ½, pp. 37-79.

16. Dani, A.H. (Peshawar, 1967). "Timargarh and Gandhar Grave Culture." *Ancient Pakistan*, No. 3, pp. 3-64.

17. Silvi Antonini, C. and G. Stacul (Rome, 1972). *The Protohistoric Graveyards of Swat (Pakistan)*, 2 Vols.

18. Jarrige, J.F., M. Santoni and J.F. Enault (Paris, 1979). *Fouilles de Pirak*.

19. *I.A.R.*, 1980-81, p.56.

20. *Ibid.*, p 56.

21. *Ibid.*, p. 55.

22. Hackett, C.A. (London, 1878). "Sketch of the Geology of Ulwur." *Gazetteer of Ulwar* (P.W. Powlett), pp. 177-183.

23. Singh, R.L. (Varanasi, 1987). *India: A Regional Geography*, pp. 517-564.

24. Sharma, Y.D. and Kumar, M. (Patiala 1980). "Excavation at Ghuram" in *Proceeding of Punjab History Conference*, pp. 32-35.

25. *I.A.R.*, 1980-81, p. 51.

26. Singh, R.L. (Varanasi, 1987), *op.cit.*, p. 123.
27. Traill, G.W. (Calcutta, 1878b). "Statistical Sketch of Kumaun." *Official reports on the Province of Kumaun* (J.H. Batten), p. 13.
28. Walton, H.G. (Allahabad, 1910). *British Garhwal*, p. 9.
29. *Ibid.*, p.5.
30. Pant, C.C. and Mishra, D.K. (New Delhi), 1983). "Mineral Resources of Kumaon Himalaya." *The Himalaya: Nature, Man and Culture* (O.P. Singh), p. 179.
31. *I.A.R.* 1982-83, p. 95.
32. Nautiyal, K.P., Khanduri, K.P., Nautiyal V. and Rajput, D.L. (Delhi, 1986). "Painted Grey Ware Culture in Garhwal, Himalaya: New Evidence and Interpretation." *Puratattva* No. 17, p. 11-14.
33. *I.A.R.*, 1983-84, p. 46.
34. *I.A.R.* 1961-62, p. 42.
35. *I.A.R.*, 1982-83, p. 16.
36. *Ibid.*, 1973-74, p. 9.
37. *Ibid.*, 1967-68, p. 9.
38. Mian Amzad Hossain, A.K.M. (Peshawar, 1969). "New Archaeological Discoveries in East Pakistan." *Ancient Pakistan*, No. 6, p. 215-22.
39. *I.A.R.* 1971-72, p. 48; *Man and Environment* No. 6, p. 111-49.
40. *I.A.R.*; 1963-64, p. 66.
41. Chakrabarti D.K. and Hassa, S.K. (Delhi, 1982). "The Sequence of Bahiri (Chandra Hazar Danga) District Birbhum, West Bengal." *Man and Environment*, Vol. VI, III, p. 49.
42. O'Malley, L.S.S. (Calcutta, 1910). *Bihar District Gazetteer Series Singhbhum*, p. 15.
43. Hunter, W.W. (1876, rpt. Delhi 1973). *A Statistical Account of Bengal*, Part IV, p. 318.
44. *Ibid.*, p. 211.
45. Jacobson, J. (Weisbaden, 1975). "Static Sites and Peripatetic Peoples: Notes on the Archaeology of Population Mobility in Eastern Malwa." *Pastoralists and Nomads in South Asia* (L.S. Leshnik and G.D. Sontheimer), p. 68-91.

46. Ball, V. (Calcutta, 1881). *A Manual of Geology of India* (Part III, Economic Geology), p. 383.
47. Krishna, V.S. (Bhopal, 1967), *Madhya Pradesh District Gazetteers Sagar*, p. 12.
48. Grant, C. (1874, rpt. Delhi, 1984). *The Gazetteer of the Central Provinces of India*, p. 436.
49. Russell, R.V. (Allahabad, 1908). *Central Province District Gazetteers.* Nimar District, *p. 147.*
50. Pandey, R.P. (Delhi, 1998). "Iron Technology: Its Role in the Historical Development in the Morar Valley, M.P., Tripathi Vibha (ed.), *Archaeometallurgy in India*, p. 58.
51. Palande, M.R. (Ahmedabad, 1961). *Gujarat State Gazetteer Broach District*, p. 13.
52. Ball, V. (Calcutta, 1981), *op.cit.*, p. 399.
53. Hegde, K.T.M. (Baroda, 1975). *Iron Metallurgy, Excavation at Dhatava* R.N. Mehta and S.N. Chowdhary), pp. 49-50.
54. Deo, S.B. & Jamkhedkar, A.P. (Bombay, 1982). *Naikund Excavations (1978-80)*, p.11.
55. Deo, S.B. (Aurangabad, 1974). *Excavations at Bhokradan*, p. 166.
56. Sankalia, H.D. and Deo, S.B. (Poona, 1960). *From History to Pre-history at Newara*, p. 383.
57. Grant, C. (1874, rpt. Delhi, 1984). *The Gazetteer of Central Provinces of India*, p. 330.
58. Allchin, F.R. (Hyderabad, 1960). *Piklihal Excavations Andhra Pradesh Government Archaeological Series No. 1*, p. 137.
59. *Geological Survey of India Miscellaneous Publication No. 30*. (Delhi, 1974), pp. 19-20.
60. Abhishankar, K. (Bangalore, 1968). *Mysore State Gazetteer Kolar District*, p. 16.
61. Thapar, B.K. (Delhi, 1952). "Porkalam 1948 Excavation of a Megalithic Urn-burial. *Ancient India*, No. 8, pp. 8-16.
62. Buchanan, F. (London, 1807). *A Journey from Madras... the Countries of Mysore, Canara and Malabar*, p. 436.
63. *Geological Survey of India Miscellaneous Publication No. 30* (Delhi, 1974), pp. 29-31.
64. *op.cit.*, pp. 20-21.
65. *I.A.R.*, 1976-77, p. 47.

66. Lechnik, L.S. (Weisbaden, 1974). *South Indian Megalithic Burials*, p. 40.

67. *I.A.R.*, 1976-77, p. 47.

68. The oxidation number is the oxidation state of an element in a compound that is equal to the number of electrons that need to be added to element in its combined state to produce the netural atom. If the oxidation number of an element increases in a reaction, the element is oxidized, if it decreases, the element is reduced. It also depicts the reactivity of an element.

69. Gangue is the minerals of little economic value which are associated with metallic ores like silica etc.

70. Blast furnace is a cylindrical, slightly conical with a vertical axis in which ore, solid fuel and slag forming materials or flux are added at the top, an air blast is admitted (through nozzles called tuyeres) near the bottom, and metal and slag can be continuously or intermittently withdrawn from the bottom. Blasting in the furnace increases its temperature in a directly proportionate manner, subject to the condition of fuel. The furnaces must have temperature above than 300-400^{0}C, but equipment of lower temperatures (less than 300^{0}C) is called oven.

71. Pig iron is the first stage of iron that has come from the blast furnace and contains too much carbon (typically 4% by mass) to be of use, being brittle. It is named after the moulds called 'pigs', used to cool the molten metal.

72. Mishra, A.K. (Delhi, 2003). *Ethno-archaeometallurgy of Iron in India*, pp. 69-93.

73. Hegde, K.T.M. (London, 1973), "A Model for Understanding Ancient Indian Iron Metallurgy, *Man*, Vol. 08, No. 3, September, 1973, p. 417.

74. Mishra, A.K. (Delhi, 2003). *Op.cit.*, p. 48.

75. Smelting is a general term to describe fire-refining of ore or concentrate to obtain metal. Reduction of oxides is usually effected by the presence of carbon or carbon monoxide, and a flux is added to form fusible and liquid slag which will float on the metal and collect the impurities which it is desired to remove.

76. Mishra, A.K. (Delhi, 2003). *Op.cit.*, p. 92.

77. Equilibrium diagram is a type of plane diagram setting out the equilibrium relationship between composition, temperature and structure. Performing the function of a metallurgist's map, they must be interpreted with caution since some alloy system require considerable time to achieve equilibrium, and commercial heat treatments usually continuous cooling effects.

78. The compound Fe_3C which is formed in steels by the process of cementation in which wrought iron bars were packed in charcoal, sealed with clay and heated to 700-900^{0}C for several days in order to diffuse sufficient carbon into the iron to convert into steel. It is a process of solid-state diffusion by which iron absorb small amount of carbon.

79. Tensile strength is the stress in a material that is being stretched. It is the tension per unit cross-sectional area.

80. Ghosh, A.K. and Chattopadhyay, P.K. 1982. *MASCA*, 2, 2, P. 63.

81. Ferrite is the body-centered cubic (alpha) form of iron, which is suitable below 760°C in the pure form. In commercial steels, many elements enter into solid solution in the ferrite, so that the term is applied to any solid solution based upon alpha iron.

82. Pearlite is the lameller carbide-containing constituent present in annealed and normalised carbon steels and in grey cast-iron, and characterised by the 'finger-print' like appearance of the alternate plates of iron and iron-carbide in the microstructure.

83. Hadfield, R. (Varanasi, 1975). *Taxila* by Marshal, J., Vol. 2.

84. Hyper-eutectic alloy means one containing more than the eutectic composition and structure which has always the lowest melting point alloy of any given alloy system, whether binary, ternary etc. and melts at a fixed temperature.

85. Anosoft reproduced it with the help of central Asian artisans at his factory in Zlatnost (the Urals). Another method was reported by Marsalaski from the Bukhara region.

86. Crystallography is the study of the structure of crystals. Crystallographic techniques can determine the Bravais lattice and the magnitudes of the lattice vectors. In X-ray crystallography the X-ray diffraction to investigate the structure of crystals by identifying the possible angles at which X-rays can be diffracted using.

CHAPTER 5

NON-FERROUS METALLURGY

In the earth's crust predominant metal is iron, whose symbol is Fe derived from the Latin word *ferrum indicum*, which is the most important metallic element constituting nearly 90% of all the metal that we use while all other metals like copper, zinc, tin, silver, gold, lead etc. which are minor in availability and in uses, are grouped under non-ferrous metals. Chemically Fe is in group d-5 while Cu, Ag and Au are in group d-9, Zn in group d-10, Sn and Pb are in group p-2 of the Periodic table means iron (ferrous) have different in chemical property, then other metals of different group like copper, silver, gold, zinc, tin and lead. Here group in the periodic table, as per periodic law, means elements with similar electronic configuration having similar chemistry. The publication of Irfan Habib's Atlas[1] and the Cambridge Economic History of India (Vol. I)[2] has facilitated the location of most important medieval ore-sites in the non-ferrous category. In non-ferrous metal lead and zinc ores commonly occur together in nature as their sulphides i.e. galena in the form of lead sulphide (PbS) and sphalerite in the form of zinc sulphide (ZnS), accounting for most of the world's production. The other minerals of these two metals are Cerussite ($PbCO_3$), Angelesite

($PbSO_4$), Simthsonite ($ZnCO_3$) and Hemimorphite ($Zn_2SiO_3(OH)_2$). Geologically the deposits of lead and zinc are either of hydrothermal (deposit precipitated from a hot mineral around igneous mass) or of sedimentary origin (formed by the deposition of sediments). Some of the important by-product of these deposits are silver, gold and cadmium.

The estimated ore reserves of lead and zinc vary from 240 to 420 million tonnes are considered recoverable in India, out of which 90% of the reserves are in Rajasthan. Barring Rampura-Agucha, all others are not of world class having metal content is approximately 1.4% Pb and 4.0% Zn. Saragipalli ore reserves in Orissa contain only lead. Of the various deposits, only two viz. Zawarwala and Ramapura-Agucha of Rajasthan are the ores which can sustain production of zinc and lead in modern time too. Deposits of lead and zinc ores occur in Jaipur and Mewar districts of Rajasthan. The deposits at Zawar ($24^0 21'$; $73^0 44'$) near Udaipur have been found to be extensive and of large size. Though it is commonly believed that pyrometallurgy of zinc at Zawar was in vogue in the 13th century A.D. but recent discoveries do established the primacy of ancient India in the field of zinc and brass metallurgy and our country was the first to make distilled zinc way back to 4th century B.C.E.[3] Scientific study of ancient Indian mines clearly point to mining activities (for argentiferous galena) at Rajpura-Dariba in Rajasthan in 1260 B.C.E. having low ore grade of lead and zinc reserves. The technology transfer of zinc distillation practices from Zawar (India) to Britol (England) took place in the early part of 18th century. At Zawar[4], lead and zinc occur as replacement veins and fissure fillings occupying fault and fracture zones in dolomites of Aravalli age associated with phyllites and quartzite. The rocks in the Mochia and Magra hills dip steeply towards north while the lodes which transverse them have a steep southerly dip. In addition to massive lodes and veins, there are fairly rich zones of dissemination in the dolomite[5], which may be workable. The ores consist mainly of a mixture of galena (PbS) and sphalerite (ZnS), the proportion of which varies greatly. It would appear that the ore is richer in lead in the upper horizons and zinc in the lower horizons. The average grade is 4% to 5% of the combined metals but richer ores occur which can yield 10% to 15% of the two metals together. The deposits are expeced to contain extensive resources running into several million tons of rather low grade. S.W.K. Morgan in 1976 have drawn the attention of archaeometallurgists of British Museum and had done a multidisplary archaeological investigation to uncover the mysteries of ancient zinc extraction technology practiced at Zawar. The investigations commenced in 1983 by Mr. H.V. Paliwal of Hindustan Zinc Limited, unearthing of two sets of ancient zinc smelting furnaces still intact with their full charge of 36 retorts turned out to be the most significant aid to decode the technology of zinc smelting during the investigations. Subsequently, investigations were carried out in Rajasthan at Dariba and Agucha in 1986-87. The extensive mine workings, heaps of mined out debries, slag, furnace refractories and large timber revetments are also found at Dariba (Dt. Rajsamand) and Agucha (Dt. Bhilwara) in Rajasthan. At Dariba, Hindustan Zinc Limited has obtained the C^{14} dates which show that the industry was in operation in *circa* 1000 BCE (3040 ± 150 B.P.). There are also several small occurrences of lead-zinc ore veins in the schistose[6] rocks of Bihar, Orissa and Madhya Pradesh but they are not exploited during ancient time.

Zinc is soft bluish-metal of transition element having a density of 7.14 g/ml and melting point is 419°C, with a close-packed hexagonal structure. Like melting point, its boiling point is also low i.e. 907°C. Its low boiling point play a very important role in its extraction from the ore. Zinc was the last common metals to be smelted in early times because it oxidizes very rapidly and had to be distilled (b.p. 907°C) in an air-tight reducing environment by the process of distillation. The first large scale production probably commenced in India rather less than a thousand years ago, but its use as alloy in Baran took place as early as 9th-10th century A.D. in the ancient Gujarat. From the critical technical study of metal images discovered from Mahudi (Dt. Mehsana), Lilvadeva (Dt. Panchmahal), Akota (Dt. Vadodara) in Gujarat, which are dated from 6th – 14th century A.D. witnessed continuous activities in the manufacturing of brass metallic images in which zinc was used as an alloy with copper, but its evolved phase at Akota was seen during 11th century A.D. The delay in production of zinc in contrast to copper and iron is because of the great difficulty in extracting it from its ores. Not only does it require extreme reducing conditions, but it is produced as vapour due to its low boiling point (907°C), which is less to the melting point of copper and iron. Thus some sort of system was necessary where the zinc was produced under extreme reducing conditions and the vapour condenced with oxygen rigorously excluded. Here zinc in the solid state as in its ore give off vapours, consisting of atoms or molecules of which have evaporated from the condensed form at a particular liquid pressure of the atmosphere. If the substance is in an enclosed space, the vapour pressure will reach an equilibrium value that depends only on the nature of the substance and the temperature. This maximum value occurs when there is a dynamic equilibrium between atoms or molecules escaping from the liquid or solid and those that strike the surface of the liquid or solid and return to it. The vapour is then said to be a saturated vapour and the pressure it exerts is the saturated vapour pressure. Thus obtaining a saturated vapour of zinc at a particular saturated vapour pressure is the most complicated technical aspect of its extraction. Thus some sort of system was necessary where the zinc was produced under extreme reducing conditions and the vapour condensed with oxygen rigorously excluded. However, brass, the alloy of copper and zinc was in regular use over 2000 years ago in India. This was made directly on an industrial scale by reacting calcinied zinc ore with charcoal and finally divided copper in a closed crucible at about 1000°C, known as cementation process[7] in which both zinc and copper diffused through the surface. In Greek, brass was called as "Oreichalkos" which was an exotic material usually originating well to the east suggesting the process began somewhere between Greece and India sometime in the 1st millennium B.C.E. although the earliest regular series of dated brass are the brass coins from Phrygia[8] and Bithynia in Antolia dating to the beginning of the 1st century B.C.E. India and Middle East had at best source of zinc and copper, and thus it is the area where one would expect brass to have been developed on large scale. In the west the cementation process became the standard way of producing brass up to the 19th century and metallic zinc was almost unknown until introduced from the east from 16th century A.D. onwards. In India the situation developed in a very different manner. The few analyses which have been performed, such as on the objects from Taxila in Pakistan suggest brass was in use from *circa* 400 B.C.E., and at least some of the pieces may have been made by adding metallic zinc to copper as their zinc content is analysed to be 34.34% zinc. In typical brass, normally copper is 63% and rest of 37% is zinc, because brass is basically an alloy of copper and zinc, containing more than 50% copper. However, it is not until about a thousand years ago that detailed descriptions of the preparation of zinc by distillation[9] appear and shortly after that there is the earliest field evidence that from Zawar for the industrial production of zinc.

The C-14 dates from Zawarmala and Mochia at Zawar confirmed the early mining operation of zinc were carried out during the later part of the 1st millennium B.C.E. The one extracted from these areas was predominantly sphalerite (ZnS), with small amounts of lead, having low silver content. This is based on the analysis of the ore. But in lower strata silver content is not low, as on the upper strata of Agucha. The heap of beneficiation debris where the ore was crushed and the mineral picked out are very low in zinc, as well as lead, although it is of course possible that they could have been reworked more recently, but for which there is no evidence. The slag heaps for the production of lead at Zawar range in date over the last 2000 years. However, excavations of slag heaps at Zawar significantly has not produced a single piece of cupellation debris as reported by its excavators P.T. Craddock and K.T.M. Hegde. Thus there is evidence for zinc mining on a massive scale and lead smelting on a minor scale 2000 years ago without any evidence of contemporary zinc smelting. One possibility is that the earlier evidence could be buried beneath the later heaps, but if so it was possible probably on a minor scale. It is more likely that the sphalerite (Za, Fe)S, zinc blende (containing some iron) was taken elsewhere to produce brass directly by the cementation process. Although there is some evidence for the existence of metallic zinc in India at this time it seems likely that the bulk of brass would have been produced by the cementation process. Thus it would seem that the mines at Zawar were always worked predominantly for zinc, with lead recovered as a by-product. At Zawar there are enormous heaps of debris, metres thick and covering many hectares made up of countless millions of clayretrots in which precipitated alloy deposited from vapour. Each retrot was used but once and then discarded and the great heaps must have built up quickly, frequently burying the blocks of furnace in which the process took place. From a detailed examination of the retrots and furnaces linked with the contemporary descriptions it has been possible to reconstruct the process of zinc smelting in some detail. An examination of clay retorts at Zawar shows that they belong to period to the 14th century A.D. and contain carbonaceous reducing agents. They were found with the pure zinc metal condensed in the neck of the retorts.[10] On the other hand, Forbes grudging remark that "though the value of the old Indian alchemists and their modern commentators is very doubtful, it seems that zinc was prepared by Indian chemists since the 12th century A.D. (why not 10th century?) but that this remained a laboratory experiment and never was applied to industrial production" is an example of Eurocentric bias, which prior to Needham's revealing researches on China, was a marked feature in the history of metal technology.

The process of ancient Indian zinc metallurgy is based on Zawar. The ore at Zawar is basically sphalerite with some marmatite[12], pyrite and galena, all sulphides. Before any reduction could take place it was essential to remove both the lead and sulphur. The galena (PbS) is macroscopically separated from the sphalerite, and could be separated manually, and the remainder of the ore would have been roasted in air to convert sulphide to oxide by the process of oxidation. Even a few percent of residual sulphides would severely inhibit the reduction process. Thus it is likely that the roasting must have been carefully carried to completion in special built ovens, but no remains of these have so far been found at Zawar. But retort was found in which calcined ore was readily reduced into zinc metal. The form of the retort was an open ended clay vessel. After filling, an open ended funnel-shaped clay condenser was filled over the open end and sealed in place with clay. In this retort following reaction was taking place as

$$ZnO + CO \rightarrow Zn + CO_2$$

The aforesaid reaction is strongly endothermic[13] which required a large heat input from outside, through the retort walls and the charge itself. Conduction is quite slow and if it was the main means of heat transference the reaction would have proceeded unequally through the charge. However, with an open structure then the heat transfer was instantaneous by radiation. Also the reaction is largely governed by the gases moment of both the CO needed to bring about the reduction, and the zinc vapour to escape. Without an open structure this could not happen, and in a tightly packed retort the reaction would be unable to take place. In the above reaction locally available salt was used as fluxes which promote the sintering of the dolomite preserving the open structure. The carbon monoxide (CO) generated from the organics and charcoal would reduce the zinc oxide to zinc vapour which could escape to the condenser down the central passage left by the now burnt stick. It was very essential to maintain very reducing conditions as the reaction is easily reversible unless the CO_2 had to be removed or reduced into CO. The proportion of the charge and the temperature were also critical, too much flux or too hot and the charge would melt, but too little flux or too cool and the charge would just fall out to the retort as a dry powder. Cow dung was widely used more recently in India both as a binder and fuel, and it is likely that this was also the case at Zawar. Thus the large retort would have contained about a kg. Of calcide ore and dolomite with charcoal and about 1% of salt bound together by cow dung in balls about 1 or 2 cm in diameter. Excavation at Zawar, the furnaces are of industrial purposes (large amount of zinc production) because each held 36 retorts in a six by six arrangement and the furnace were themselves arranged in banks. At Zawarmat a seven furnaces were excavated in which no less than 252 retorts were fired simultaneously. The retorts were inverted in the furnace with the condenser necks protruding through the perforated bricks which separate the furnace from the cool chamber beneath. Each furnace had four perforated supported on a ledge running round the furnace walls and central pillar in the cool chamber. Thus perforated bricks separated the furnace from the cool chamber. Each of the perforated bricks had nine large holes to accommodate the condenser necks and smaller holes for air and ash to pass. The forming zinc vapour condensed and dripped out of the condenser ends in the cool chamber below to be collected in vessels placed beneath. Under ideal conditions zinc vapour condenses at 913°C and solidifies at 419°C then vapour condenses as a product known as blue powder which is made up of minute globules of zinc coated with zinc oxide and is completely useless. It was impossible to suppress it totally, and at Zawar too, the condenser necks always have some blue powder present. As mentioned above the zinc vapour had to be kept in an intensely reducing condition of carbon monoxide, which burnt off at the condenser ends. Study of the degree and depth of vitrification of the retort and furnace walls by British scholar I.C. Freestone and Lynn Willes under scanning microscope enables some estimate of the temperature and duration of the process to be made. The temperature as calculated from the retorts was in the range of 1150 – 1250°C which was low compared to the temperature of 1400°C used in the modern retort process. The depth of vitrification on the furnace walls suggest firing times between four to six hours. Combined scane electron microscope and X-ray diffraction study show the conditions to have been extremely reducing with the partial pressure[14] of oxygen was upto 10 or less. Thus at Zawar, the technical sophistication and application of scientific principles in zinc production are unparalleled elsewhere in the late-Medieval period, and the element of standardization and mass production foreshadow the Industrial Revolution. This is the earliest example of high temperature distillation under optimum reducing condition.

India has a very old and fascinating history of non-ferrous metal like gold. Out of all metal known to man in yore, perhaps gold was the first which man became acquainted, due to its pleasing and untarnishing colour, excellent corrosion and oxidation resistance, ease of forming with less technical complexity in its extraction. Its melting point is 1064.4°C and density is 19.3. Gold occurs native (Native gold has a purity in excess of 85%, is always contaminated with silver and copper), typically as small particles in quartz but also as deposits after erosion of the quartz and sometimes as nuggets.[15] Gold-bearing quartz veins occur in shear zones in the schistose Dharwar and near the margin of the Dharwars and gneisses in the region of Andhra Pradesh and Karnataka. Its characteristics also made it a symbol of sacredness in ancient India. The attributes of gold influenced the mind and heart of early Indians so deeply that they conferred upon the supreme spirit the designation of *Hiranyagarbh* means he remains in a golden egg as an embryo. Most of the trade practices indulged in by the medieval Indian merchants can be termed as 'mercantilist' because they invariably involved an inward flow of gold and silver. The principal sources of the gold in India is the Kolar gold field in Kolar District of Karnataka which occurs in the area of 65 km in length and 4 to 6 km in width. It is occupied by metamorphosed and folded mafic rocks[16] and silts of associated basic intrusives. Some gneissic and ferruginous rocks occur along the margins. The auriferous lodes of the Kolar gold-field are contained in the quartz veins, which run parallel to one another in the north-south direction in a belt of hornblended schists along shear zones. The gold is associated with pyrite, pyrrhotite and arsenopyrite, the ore being hypothermal in origin means deposit formed at a great depth and high temperature ranges from 300°C – 500°C. The most productive of these is a single quartz-vein, about 1.2 meter thick, which bears gold in minute particles. Mining operation in this reef have been carried to a depth of 3050 meter at present thus it was not exploited in the yore. At some other places in the Peninsula, the gold is revealed by many signs of ancient gold-working in diggings, heaps of crushed quartz, stone-mortars, have been reported like Kolar and these are specially in Hatti. Hatti gold fields present in the Maski band of Dharwarian schists, in the Raichur district of Karnataka. In Wainad region of Nilgiri district, numerous gold bearing quartz lodes were opened up and worked between

1880 and 1900 A.D. In the Anantapur district of Andhra Pradesh, near Ramagiri, some quartz reefs were worked between 1908 and 1924. Other areas where gold-quartz veins are known are in Jashpur and Dhalbhum and Singhbhum in Jharkhand and Sambalpur and Koraput district in Orissa.

In contrast to quartz-vein gold, the alluvial gold as 'placer' metal in India is much wider. Many of the rivers draining the crystaline and metamorphic tracts in India are reported to have auriferous sands, but only a few of them contain gold in a sufficient quantity to pay any commercial attempt for its extraction in modern time but they must be exploited in the past. The only instance of successful exploitation of this kind is the dredging of the Upper Irrawaddy, in search of the gold bearing gravel in its bed, for some years; but the returns fell off and operation were closed down in 1918. Alluvial gold-washing is carried out in the sands and gravels of many rivers of Madhya Pradesh and in sections of the Indus Valley at Ladakh, Baltistan, Gilgit and Attock in India and Pakistan, but none of them are of any richness. The quantity won by the indigent workers is just enough to give them their day's wages with only occasional windfalls.

Alluvial gold of Upper Indus plains has been known to be extracted in the past. In Attock district of Punjab in Pakistan, gold is found on the banks of the Indus river. In the nineteenth century the right to extract gold was granted yearly to a contract rate in 1884 being Rs. 120. Apart from the Indus, gold was found on the banks of other streams like the Raish, Sil and Sohan in the same district of Attock. In Rawalpindi, gold is found in the tributaries of the Jhelum and of the Indus. In Jhelum district gold is found in the shape of minute scale in salt range and in the beds of the numerous nallahs which flow through the miocene formations, the sand is washed for gold. It seems to be obtained in the largest quantity towards the Indus, north of salt range. In the same district of Jhelum in upper Indus plain, the rivers where alluvial gold is noteworthy are the Bunhar and the Kahari. The Kahari river contained gold in its course below Rohtas and Bakrala. In Jammu and Kashmir, the occurrences of gold have been noted in Baramulla and Ladakh districts. In Baramulla, the gold is present in traces at Lashteal hill spur, while in Ladakh alluvial gold is seen in Kio area where concentrations reached 6.65 grams per cubic yard of gravel.

In Rajasthan auriferous gold (gold bearing mineral deposit of iron) were noticed by Major Hughes near Rohri in 1897. Recently gold mineralization is certain area near Rohri, as Jahazpur belt is reported by R.S. Jain in 1987. In Bhilwara district at Devtalai and Amargarh contain indications of gold mineralization. The core sample of quartz reef from Devatalai assayed 0.2 gm Au/tonne while the surface grab sample of sheared quartz reef from the pit mouth of one of the old workings in Amargarh assayed 1.8 gm Au/tonne. In Ajmer district about 3 km east of Kharwa, south-east of a small village called Gawara, is well which is reported to be an old working for gold. In Jhunjhunu district in the Kolihan copper mine, ore shoots with high copper content were also found to contain high percentage of gold 0.4% to 1.53%. At Usri in the same district of Jhunjhunu, a large number of old workings in the south-western part were supposed to have been worked for gold. In Gujarat, gold is reported from Varodara and Jamnagar districts. In Varodara, gold occur in association with Galena near Barkunda in which 3% lead contained 7 ppm of gold. At Jamnagar, the Alech hills south-east of Satapur apparently contained gold. In Maharashtra gold is reported from two districts i.e. Bhandara and Nagpur. The Geological Survey of India have found three auriferous quartz veins at Kolari, Mokhanari and Pular near Bhiwapur in Bhandara district. In the same district alluvial gold is reported in the Maru stream at Ambagarh, near Thirora on a Tributary of the Wainganga and in bed of the Son Dhodha stream. In Nagpur district some places gold from the sands of a small stream called 'Dabari Pat' at a place 1.6 km north of Pular is said to have been extracted. The Sonajhari people are also known to pan the sand of the Maru river and its tributaries after the monsoons for gold.

In chalcolithic Harappan culture, gold has been used extensively for making ornaments. The gold used is rather pale in colour and it has been suggested that quite often it was used as electrum which is a natural alloy of gold (55% to 85%) and silver. Considering that gold from the Kolar and

Hatti mines of Karnataka is also of light colour like electrum, it is justifiable to infer that surface collection of gold nuggets had started by the Harappan period. Clustering of the neolithic sites around gold mining areas in Karnataka and recovery of gold ornaments from Tekkalkota[17] (15°32'; 76°53') and Daimabad[18] (19°31'; 71°05') confirm that the exploitation of these gold mines was contained, even after the Harappan period. The C^{14} dates confirm that the tradition of gold extraction in Karnataka started in the first millennium B.C.E. At Hatti in Karnataka, a narrow reef had been worked by the ancients to a length of 1500 meter and to a depth of 190 meter by fire-setting alone.[19] The PRL-253 C^{14} sample from Hatti has been dated as 760 ± 150 BCE. During the pre-Christian era, Hatti and Maski in the Raichur doab of Karnataka were the main gold working mines. The neolithic sites of Piklihal, Kotekal, Maski etc. cluster around the Maski gold field on which we find Asoka's Minor Rock Edicts (three out of five rock edicts; the other two are on the neighbouring diamond field of Yerragudi and Pattikonda). The Mauryan association with Karnataka was evidently related to the state demands of gold and diamonds. Hatti continued to yield gold in the post-Christian era confirmed by C^{14} dates of 30 to 110 A.D. in Oakley's 80 meter depth shaft. At Kolar champion reef at 50 meter depth was reached at 475 A.D. if not earlier confirmed by C^{14} date. A later date of 685 A.D. from a superficial alyer at Kolar suggests difficulty with the hard rock retreat mining. In northern Karnataka the evidence of old gold workings were seen in Chittradurga, Gulburga, Dharwar, Shinoga and Raichur districts and in southern Karnataka important gold deposits are also found in Kolar district. But it is important to note that in northern Karnataka, gold artifact in Neolithic phase found only at Tekkalakota, means this precious metal seems to have become more significant in the life of the region in early historical phase. In Tamil Nadu the gold deposits have been seen in Coimbatore district, where the noteworthy deposits occur at Bensibetta, Porsedyke mine, Hadabanatta and Kavudhalli. Old working are seen at Hadabanatta and Kavandhalli. In the same state of Tamil Nadu 750 square kilimeters of gold deposit is seen in the Wynad-Malabar area, lying to the West of Nilgiri mountains in which gold reef and placer deposits have been found. The noteworthy regions are Nadghani-Devala-Pandalur area, Chulimala, Attikuna and Richmond. Most of the above-mentioned areas are also worked by old workings in the form of pits and shafts.

Thus gold like copper, was one of the earliest metals to be exploited by man due to it natural occurrence on nuggets of native metal. The most common source of ancient gold would have been nuggets or fine particles of native metal in alluvial deposits and not the ore. Native gold has a purity in excess of 85% having alloy elements like silver, copper mercury and bismuth. It is the most malleable metal known, and may be beaten into leaf $.5 \times 10^{-7}$ cm in thick, which transmits green light. Having a fine yellow colour, the metal has a density of 19.3 gm/ml, a melting point of 1063°C, an annealed hardness of $20H_D/30$, a tensile strength of about 7 tons/cm^2. Albeit gold is an element of the copper group electronic configuration shows oxidation states +5, +3, +2 and +1, but it is less oxidised in the air having highly corrosion-resistant by which it was originally used for ornaments and coinage. Although its use for coinage has now ceased but it remains the international standard for world finance. Gold is not attacked by environmental corrosion or single acid (except H_2SO_4). It dissolve only in acqua ragia (25% nitric acid and 75% hydrochloric acid) and in halogen solutions and it is attacked by flourine. Due to this noble character and low hardness result in an ability to pressure weld at low temperature, which renders it very suitable for manufacturing of ornaments. Since native gold is usually occurs alloyed with other metals thus to make pure gold, these impurities have to be removed by a complex refining process. In any case, the remarkable malleability of pure gold (which allows it to be cold worked[20] without annealing[21]) would usually be undesirable in artifacts. The difficulty can be overcome by alloying with silver or copper. The purity of the resulting alloys is described in carates (commonly 9, 18 and 22) where these are the number of parts of gold (by weight) out of 24 parts. This is because gold forms suitable metal – metal bonds with other metals like copper and silver. The metallurgical extraction of gold is very complicated, thus in ancient period it was extracted as a by product in the extraction of copper and lead.

Notes and References

1. Habib. I. (Delhi, 1982). *Atlas of Mughal Empire.*
2. More specifically, see Irfan Habib's chapter on "Monetary System and Prices," Vol. I, pp. 360-77.
3. C^{14} dates from Zawar mines are 380 ± 45 BCE; 360 – 280 BCE; 540 – 385 BCE; 375 ± 40 BCE; from Mochia 410 – 390 BCE; from Agucha 405 – 390 BCE; 390 – 375 BCE and from Dariba mines 400 – 190 BCE and 370 ± 45 BCE.
4. The Aravalli hills run over 1000 kms. approximately northeast from Gujarat to Delhi principally through the state of Rajasthan. Although the hills are rarely mountainous they are ranged in a series of ridges running parallel rather in the manner of a grid iron and in historic times at least they have formed a considerable barrier to communication. These ridges were formed by the repeated isoclinal folding of the highly metamorphosed Precambrian rocks. Seven principal shear zones have been located in Aravallis and all are rich in minerals, the northeast zone tends to reach in copper (Khetri etc.), whereas the southwest zone is rich in lead-zinc ores (the mines of Ambaji, just inside the state of Gujarat); although exploited quantities of copper, zinc and lead are found throughout.
5. Chemically dolomite is $CaMg(CO_3)_2$ occurring as hexagonal, rhombohedral crystals with curved surface. It is sedimentary rocks of secondary origin.
6. Shistone pertaining to rocks of cleavage characteristic of metamorphic rocks notably schists and phyllites, in which rocks tend to split along parallel planes defined by the distribution and parallel arrangements of platy crystals.
7. Cementation is a metallurgical process in which the surface of a metal is impregnated by some other substance, especially an obsolete process for making steel by heating bars of wrought iron to red heat for several days in a bed of charcoal. Cementite which is a resultant product of cementation in iron is Fe_3C.
8. Phrygia was a short-lived kingdom of the west central Turkey; it lasted from *circa* 750 – 680 BCE when it was destroyed by the Cimmerians. The capital of the Phrygian kingdom was at Gordion I is known also from the rock-cut sanctuaries and burial remains, including rich burials under large tumuli. The Phrygians are

thought to have moved into Antolia from thrace or Macedonia and to represent an Indo-European aristocracy, dominating an indigenous poplation.

9. Distillation is the process of separation of a liquid from a solid or another liquid by vaporization followed by condensation. The distillation can be carried out at atmospheric pressure or reduced pressure.

10. J. Mehmood, "Zinc Metallorgy of Medieval India," paper read at a Seminar on Science and Technology in India (1400-1800) I.N.S.A., New Delhi, 1978.

11. Forbes, *Studies in Ancient Technology*, Leiden, Vol. III, p. 281.

12. Marmatite is a dark brown to black mineral composed of iron-bearing spablerite and it is also known as Christophite.

13. Endothermic is a reaction in which heat is absorbed. According to the laws of thermodynamics a spontaneous endothermic reaction must be accompanied by an increase in entropy of the system and surroundings.

14. Partial pressure is in a mixture of gases or vapours each constituent can be considered to contribute to the total pressure that pressure which the same amount of the constituent would exert if it sere present alone in a vessel of the same volume as that occupied by the mixture. This is called the partial pressure of the constituent.

15. Nugget is a small mass of metal found free in nature.

16. Mafic rocks are a type of igneous rock.

17. Tekkalakotta is a predominantly neolithic site in Dt. Bellary of Karnataka from C^{14} dates ranges between 1540 ± 100 BCE and 1780 ± 105 for the early phase and 1610 ± 140 BCE for the later phase of neolithic. In early phase coiled ear pendent of gold is reported.

18. Daimabad in Dt. Ahmednagar of Maharashtra is a neolithic-chalcolithic site whose neolithic phase is similar to the Brahmagiri I of Karnataka.

19. Allchin, F.R. 1962. "Upon the Antiquity and Methods of Gold Mining in Ancient India" in *Journal of Economic and Social History of the Orient*, Vol. V, pp. 195-211.

20. Cold-working is defined as plastic deformation of a metal at a temperature low enough to result in permanent strain – hardening. In practice, the metal must be rolled, drawn, swayed, dammered or bent etc… at a temperature well below the recrystallisation temparature (since as the degree of plastic deformation is increased, the temperature at which subsequent recrystallization can occur is

decreased). The effect of cold-work may be removed by subsequent annealing, and cold-working followed by annealing is the only method available for reducing the grain size of pure metals and many solid solutions.

21. Annealing is a heat-treatment used to soften metal. It is also used to promote homogeneity by diffusion, relieve residual stresses arising from cold work operations and may also result in grain refinement by recrystallisation. The temperature to which the metal is raised, and the time for which the temperature is maintained depend upon the metal concerned and the particular effects that are desired.

BIBLIOGRAPHY

Abhishankar, K. (1968). *Mysore State Gazetteer Kolar District*. Bangalore.

_______ (1970). *Mysore State Gazetteer Raichur District*. Bangalore.

_______ (1972). *Mysore State Gazetteer Bellary District*. Bangalore.

_______ (1975). *Karnataka State Gazetteer Shimoga District*. Bangalore.

_______ (1977). *Karnataka State Gazetteer Bidar District*. Bangalore.

Agrawal, B.D. (1978). *Rajasthan District Gazetteer Sikar*. Jaipur.

Agrawal, D.P. (1966). C14 Dates, Banas Culture and the Aryans. *Current Science*. XXXV: 114-17.

_______ (1971). *The Copper Bronze Age in India*. Delhi.

_______ (1982).*The Archaeology of India*. London.

Agrawal, D.P., Krishnamurthy, R.V. and Kusumgar, S. (1978). New Data on the Copper Hoards and the Daimabad Bronzes. *ME* II: 41-50.

Agrawal, O.P. (1983). Scientific and Technological Examination of some objects from Atranjikhera. *Excavations at Atranjikhera* (R.C. Gaur). Delhi: 487-88.

Agrawala, R.C. (1979). More Copper Finds from Rajasthan. *ME* III: 91-2.

______ (1980). Kurdi (?) Copper Hoard from Rajasthan. *ME* IV: 89-91.

______ (1981a). Recent Explorations in Rajasthan. *ME* V: 59-63.

______ (1981b). Prehistoric copper Objects from Rajasthan. *Cultural Contour of India* (V.S. Srivastava). Delhi: 25-34.

______ (1984). Aravalli, the Major Source of Copper for the Indus and Indus related cultures. *Frontiers of the Indus Civilization* (B.B. Lal and S.P. Gupta). Delhi: 156-62.

Agrawala, R.C. and Kumar, V. (1982). Ganeshwar-Jodhpura Culture: New Traits in Indian Archaeology. *Harappan Civilization* (G.L. Possehl). New Delhi: 125-34.

Ahmed, N. (1975). *Mahasthan*. Dacca.

Aitken, E.H. (1907). *Gazetteer of the Province of Sind*. Karachi.

Allchin, B. (1985). Ethnoarchaeology in South Asia. *SAA* 1983 volume I (J. Schotsmans and M. Taddei). Naples: 21-33.

Allchin B. and Allchin, R. (1982). *The Rise of Civilization in India and Pakistan*. Cambridge.

Allchin, F.R. (1960). *Piklihal Excavations. Andhra Pradesh Government Archaeological Series No. I*. Hyderabad.

______ (1961). *Utnur Excavations. Andhra Pradesh Government Archaeological Series No. 5*. Hyderabad.

______ (1982a). Antecedents of the Indus Civilization. *Proceedings of the British Academy LXVI*: 135-160.

______ (1982b). How Old is the City of Taxila? *Antiquity* 56: 8-14.

______ (1984). The Northern Limits of the Harappan Culture Zone. *Frontiers of the Indus Civilization* (B.B. Lal and S.P. Gupta). New Delhi: 51-4.

Allchin, F.R., Allchin B., Durrani, F.A. and Khan, M.F. (1986). *Lewan and the Bannu Basin*. Oxford.

Amzad Hossain Mian, A.K.M. (1969). New Archaeological Discoveries in East Pakistan, *PA* 6: 215-22.

Ansari Z.D. and Dhavalikar, M.K. (1975). *Excavations at Kayatha*. Pune.

Ansari, Z.D. and Mate, M.S. (1966). *Excavations at Dwarka*. Poona.

Antonini, C.S. (1973). More about Swat and Central Asia. *EW* 23: 235-44.

Antonini, C.S. and Stacul, G. (1972). *The Protohistoric Graveyards of Swat (Pakistan)*. Rome.

Asthana, S. (1982). Harappan Trade in Metals and Minerals: A Regional Approach. *Harappan Civilization* (G.I. Possehl). New Delhi: 271-85.

________ (1984). The Place of Shahdad inIndus-Iranian Trade. *Frontiers of the Civilization* (B.B. Lal and S.P. Gupta). Delhi: 353-62.

________ (1985). *Pre-Harappan Cultures of India and the Borderlands*. New Delhi.

Athavale, V.T. (1964-5). Chemical Analysis and Metallographic Examination of Metal, Prakash, 1955: A Chalcolothic Site in the Tapti Valley. *AI* 20-1: 131-9.

Atkinson, T.W. (1860). *Travels in the Regions of the Upper And Lower Amoor* ... London.

________ (1875). *Statistical Descriptive and Historical Account of the North Western Provinces of India* Vol. II. Meerut Division, Allahabad.

________ (1876). *Statistical Descriptive and Historical Account of the North Western Provinces of India* Vol. IV, Agra Division. Allahabad.

Baillie, A.F. (1890, rpt. 1975). *Kurrachee Past, Present and Future*. Karachi.

Ball, V. (1881). *A Manual of the Geology of India*. Part III. Economic Geology. Calcutta.

Banerjee, N.R. (1956). The Megalithic Problem of Chingleput in the Light of Recent Exploration. *AI* 12:21-34.

_______ (1965). *The Iron Age in India*. New Delhi.

_______ (1966). Amirthamangalam 1955: A megalithic urnburial site in District Chingleput, Tamil Nadu. *AI* 22:3-36.

_______ (1986). *Nagda. MASI NO: 85*. Delhi.

Banerjee, N.C. (1925, rpt. 1980). *Economic Life and Progress in Ancient India*. Allahabad.

Begbie, I.F. and Nelson, A.E. (1909). *Central Provinces District Gazetteers Chanda District*. Allahabad.

Beglar, J.D. (1878). *Report of a Tour in Bundel Khand... Circle Report VII*. Calcutta.

Begley, V. (1986). From Iron Age to Early Historical in South Indian Archaeology. *Studies in the Archaeology of India and Pakistan* (J. Jacobson). New Delhi: 297-319.

Bevan, E.R. (1921, rpt. 1955). Alexander the Great. *The Cambridge History of India* Volume I (E.J. Rapson). Delhi: 309-46.

Bhan, K.K. (1986). Recent Explorations in the Jamnagar District of Saurashtra. *MEX*: 1-22.

Bhan, S. (1971-2). Siswal: A Pre-Harappan Site in Drishadvati Valley. *Puratattva* 5: 44-6.

_______ (1973). The Sequence and Spread of Prehistoric Cultures in the Upper Saraswati Basin. *Radiocarbon and Indian Archaeology* (D.P. Agrawal and A. Ghosh). Bombay: 252-63.

_______ (1975). *Excavation at Mitathal (1968) and other explorations in the Sutlej-Yamuna Divide*. Kurukshetra.

Bhan, S. and Shaffer, J.G. (1978). New Discoveries in the Northern Haryana. *ME* II: 59-68.

Bhardwaj, H.C. (1968-9). Rajghat Copper—A Metallurgical View. *Puratattva* 2:30-4.

_______ (1973). Aspects of Early Iron Technology in India. *Radiocarbon and Indian Archaeology* (D.P. Agrawal and A. Ghosh). Bombay: 391-400.

_______ (1981-3). Studies in Ancient Indian Technology: A Review. *Puratattva* 13 & 14: 139-48.

Bhardwaj, H.C. & Mishra, S. (1968). Early Indian Silver Coinage—A Metallurgical View. *Journal of Numismatic Society of India* 30: 241-51.

Bhattacharya, A.N., Chaturvedi, D.N., Rathore, R.S., Khan, M., Verma, L.N., Bharadwaj, D.C., Kothari, D.K., Hiran, S.L. and Kumawat, B.L. (1987). Udaipur—Gwalior Region. *India: A Regional Geography* (R.L. Singh). Varanasi: 517-64.

Bhattacharya, P.N. (1977). *Historical Geography of Madhya Pradesh From Early Records*. Delhi.

Bhoothalinga, S. (1969). *Techno-Economic Survey of Jammu & Kashmir*. New Delhi.

Biscoine, R. (1973). Dynamics of an Early South Asian Urbanization: the First Period of Shahr-i Sokhta and its Connection with Southern Turkmenia. *SAA* (N. Hammond). London: 105-18.

Bisht, R.S. (1978). Banawali: A new Harppan Site in Haryana. *ME* II: 86-8.

_______ (1987). Furthern Excavation at Banawali: 1983-84. *Archaeology and History*, volume I (B.M. Pande & B.D. Chattopadhyaya). Delhi: 135-56.

Bisht, R.S. and Asthana, S. (1979). Banawali and Some Other Recently Excavated Harappan Sites in India. *SAA 1977* (M. Taddei). Naples: 223-40.

Biswas, S.K. (1979). On Systems Theory in Anthropology. *Current Anthropology* 20: 164.

Brill, R.H. and Whampler, J.M. (1967). Isotope Studies of Ancient Lead. *American Journal of Archaeology* 71: 63-77.

Briton, D. and Richards, E.E. (1963). Optical Emission Spectroscopy and the Study of Metallurgy in the European Bronze Age. *Science in Archaeology* (D. Brothwell and E. Higgs). London: 499-509.

Brockman, D.I. (1909). *Banda A Gazetteer... of Agra and Oudh*. Allahabad.

_______ (1911). *Mirzapur a Gazetteer*. Allahabad.

Bruce, C.G. (1915, rpt. 1950). *Kashmir*. London.

Buchanan, F. (1807). *A Journey from Madras...the Countries of Mysore, Canara and Malabar*. London.

_______ (1934, rpt. 1986). *An Account of the Districts of Bihar and Patna in 1811-1812*. New Delhi.

Buller, R.H. (1906). *Baluchistan District Gazetteer Series IV & IVA Bolan and Chagai*. Karachi.

_______ (1907). *Baluchistan District Gazetteer Series VII & VIIA. Makran and Kharan*. Bombay.

Buller, R.H. and Rai, J. (1907). *Baluchistan District Gazetteers Series V. Quetta—Pishin District*. Ajmer.

Campbell, J.M. (1880a). *Gazetteer of the Bombay Presidency* Vol. XII. Khandesh. Bombay.

_______ (1880b). *Gazetteer of the Bombay Presidency* Vol. VI. Bombay.

_______ (1885). *Gazetteer of the Bombay Presidency* Vol. XVIII Poona. Bombay.

_______ (rpt. 1984). *Gazetteer of the Bombay Presidency* Vol. LXIII Thana. Pune.

Casal, J.M. (1964). *Fouilles d'Amri*. 2 Vols. Paris.

_______ (1973). Excavations at Pirak, West Pakistan. *SAA* (N. Hammond). London: 171-80.

_______ (1979). Amri: An Introduction to the History of the Indus Civilization. *Essays in Indian Protohistory* (D.P. Agrawal and D.K. Chakrabarti). Delhi: 99-112.

Chakrabarti, D.K. (1968). Origin of Ahar Culture. *Man in India* 48: 97-105.

________ (1972-3). Concept of Urban Revolution and the Indian Context. *Puratattva* 6: 27-31.

________ (1976). The Beginning of Iron in India. *Antiquity*. 114-24.

_______ (1977a). India and West Asia—An Alternative Approach. *ME* I: 25-38.

_______ (1977b). The External Trade during the Harappan Period: Evidence and Hypothesis. Paper read at seminar at Institute of Advanced Study, Shimla.

______ (1979a). Early Iron Age in the Indian Northwest. *Essays in Indian Protohistory* (D.P. Agrawal and D.K. Chakrabarti). Delhi: 347-64.

______ (1979b). Iron in Early Indian Literature. *JRAS*: 22-30.

______ (1981-83). Study of the Iron Age in India. *Puratattva* 13 & 14: 81-5.

______ (1984a). Origin of the Indus Civilization: Theories and Problems. *Frontiers of the Indus Civilization* (B.B. Lal and S.P. Gupta). New Delhi: 43-50.

______ (1984b). The Archaeology of the Iron Age and the Early Historical Period of Kashmir. *ME* VIII: 109-16.

______ (1985-6). The Pre-Industrial Mines of India. *Puratattva* 16: 65-71.

______ (1989). *Theoretical Issues in Indian Archaeology*. Delhi.

______ (1990).*The External Trade of the Indus Civilization*. Delhi.

______ (1985). A Preliminary Report on the Archaeology of Singhbhum with Notes on the Old Mines of the District. *ME* IX: 128-50.

Chakrabati, D.K. Bhattacharya, D.K. and Chattopadhyay, R.K. (1987). Metal Industries of Purulia. *ME* VII: 131-41.

Chakrabarti, D.K. and Hasan, S.J. (1982). The Sequence at Bahiri (Chandra Hazar Danga), District Birbhum, West Bengal. *ME* VI: III-49.

_________ (1988). A Preliminary Report on the Archaeology of the Ranchi District with a Note on the Asura Sites. *ME* XII: 29-54.

Chakrabarti, D.K. Nag, A.K. and Chakrabarti, S. (1978-9). An Archaeological Reconnaissance in Birbhum. *Puratattva* 10: 25-32.

Chakrabarti, H. (1966). *Trade and Commerce of Ancient India*. Calcutta.

Chapekar, B.N. (1969). *Report on the Excavations at Ter* (1958). Poona.

Charlesworth, M.P. (1924). *Routes and Commerce of the Roman Empire*. England.

Chatterjee, G.C. (1972). General Report of the Geological Survey of India for the year 1965-6. *R.G.S.I.*. 100. Delhi.

Chaturvedi, S.N. (1985). Advance of Vindhyan Neolithic and Chalcolithic Cultures to the Himalayan Terai: Excavations and Explorations in the Sarayupara Region of Uttar Pradesh. *ME* IX: 101-8.

Cleuziou, S. and Vogt, B. (1985). Tomb A at Hili North (United Arab Emirates) and its Material Connections to Southeast Iran and the Greater Indus Valley. *SAA* 1983 (J. Schotsmans and M. Taddei). Napples: 249-77.

Cobden-Ramsay, I.E.B. (1910). *Feudatory States of Orissa Bengal Gazetteers*. Calcutta.

Coggin Brown, J. (1917). A Prehistoric Copper Axe from the Kurram Agency. *ARASI* 1913-14. Calcutta: 98-185.

Coggin Brown, J. and Dey, A.K. (1955). *India's Mineral Wealth*. London.

Conybeare, H.C. Fisher, F.H. and Hewett, J.P. (1884). *Statistical, Descriptive* and Historical Account of the North Western Provincs of India Vol. VII *Farukhabad and Agra*. Allahabad.

Countenho, O. and Ramamurthy, K. (1942). A Study of Rural Settlement Patterns in Maharashtra. *Indian Geographical Journal*, XLVII: 26-40.

Crole, C.S. (1879). *Chingleput District: Madras*. Madras.

Dalal, K.F. (1980). *A Short History of Archaeological Explorations in Bikaner And Bahawalpur along the 'Lost' Saasvati river. (Indica)* (rpt.)

_____ (1981). RD 89: A New Hakra Ware Site? *ME* V: 77-86.

Dale Brockman, D.L. (1909). *Banda A Gazetteer...Agra and Oudh*. Allahabad.

_____ (1911). *Mirzapur A Gazetteer... Agra and Oudh*. Allahabad.

Dales, G.F. (1962). Harappan Outposts on the Makran Coast. *Antiquity* XXXXVI: 86-92.

_____ (1968). Of Dice and Men. *JAOS* 88: 14-23.

_____ (1979). The Balakot Project: Summary of Four Years Excavations in Pakistan. *SAA* 1977 (M. Taddei). Naples: 241-73.

_____ (1981). Reflections on Four Years of Excavations at Balakot. *Indus Civilization: New Perspectives* (A.H. Dani). Islamabad: 241-74.

Dani, A.H. (1965-6). Shaikhan Dheri Excavation. *AP* II: 17-214.

_____ (1967). Timargarha and Gandhara Grave Culture. *AP* III: 1-407.

_____ (1970-1). Excavations in the Gomal Valley. *Ap* 5: 1-177.

_____ (1983a). *Chilas*. Islamabad.

_____ (1983b). *Human Records on Karakoram Highway*. Islamabad.

_____ (1988). *Recent Archaeological Discoveries in Pakistan*. France, Tokyo.

Das, S.K. (1925, rpt. 1980). *The Economic History of Ancient India*. Allahabad.

Das Gupta, P.C. (1964). *The Excavations at Pandu Rajar Dhibi*. Calcutta.

De Cardi, B. (1964). British Expedition to Kalat 1948 & 1957. *PA* I: 20-9.

_____ (1965). Excavations and Reconnaissance in Kalat. West Pakistan. *PA* 2: 86-182.

________ (1968). Excavations at Bampur, Southern Iran: A Brief Report. *Iran* VI: 135-55.

________ (1984). Some Third and Fourth Millennium Sites in Sarawan and Jhalawan, Baluchistan, in Relation to the Mehrgarh Sequence. *SAA* 1981 (B. Allchin). Cambridge: 61-8.

________ (1989). Harappan Finds from Tomb 6 at Shimal, Ras al-Khaimah, United Arab Emirates. *SAA* 1985 (K. Friefelt and P. Sorensen). London, Riverdale: 9-14.

Deo, S.B. (1970). *Excavations at Takalaghat and Khapa*. Nagpur.

________ (1973). *Mahurjhari Excavation 1970-72*. Nagpur.

________ (1974). *Excavations at Bhokradan*. Aurangabad.

________ (1982). *Recent Researches on the Chalcolithic and Megalithic Cultures of the Deccan*. Madras.

Deo, S.B. and Ansari, Z.D. (1965). *Chalcolithic Chandoli*. Poona.

Deo, S.B. and Dhavalikar, M.K. (1968). *Paunar Excavation*. Nagpur.

Deo, S.B. and Jamkhedkar, A.P. (1982). *Naikund Excavations* (1978-80). Bombay.

Deo, S.B. and Joshi, J.P. (1972). *Pauni Excavation*. Nagpur.

Deo, S.B. & Majumdar, G.G. (1969). *Songaon Excavations*. Poona.

Deva, K. and McCown, D.E. (1949). Further Exploration in Sind: 1938. *AI* 5:12-30.

Dey, N.L. (1927, rpt. 1971). *The Geographical Dictionary of Ancient and Medieval India*. New Delhi.

Dhavalikar, M.K. (1975-6). Settlement Archaeology of Inamgaon. *Puratattva* 8: 44-54.

________ (1977). Inamgaon: The Pattern of Settlement. *ME* I: 46-51.

________ (1977-8). An Early Warehouse on the Western Coast. *Puratattva* 9: 100-3.

_______(1978). Early Farming Communities of Karnataka and Maharashtra. *Archaeology of Karnataka* (A.V.. Narasimha Murthy). Mysore: 41-7.

______ (1978-9). New Archaeology and the Indian Situation. *Puratattva* 10: 33-8.

Dhavalikar, M.K. and Ansari, Z.D. (1988). Other Artifacts. *Excavations at Inamgaon*. Pune: 554-72.

Dhavalikar, M.K., Sankalia, H.D. and Ansari, Z.D. (1988). *Excavations at Inamgaon*. Pune.

Dichter, D. (1967). *The North West Frontier of West Pakistan*. Oxford.

Dikshit, K.R. and Singh, K. (1987). Maharashtra Region. *India : A Regional Geography* (R.L. Singh). Varanasi: 698-734.

Dowie, J. (1916, rpt. 1974). *The Punjab, North-West Frontier Province and Kashmir*. Delhi.

Drew, F. (1875, rpt. 1971). *The Jammu and Kashmir Territories: A Geographical Account*. Delhi.

Dupree, I., Gouin, P. and Omer, N. (1971). The Khosh Tapa Hoard from North Afghanistan. *Archaeology* 24: 28-34.

During Caspers, E.C.L. (1970). Trucial Oman in the Third Millennium B.C. *Origini*: 205-76.

Durrani, F.A. (1981). Evidence West of Indus. *Indus Civilization New Perspectives* (A.H. Dani). Islamabad: 133-8.

________ (1984). Some Early Harappan Sites in Gomal and Bannu Villages. *Frontiers of the Indus Civilization* (B.B. Lal and S.P. Gupta). New Delhi: 505-10.

Dyson, R.H. (1965). Problems in the Relative Chronology of Iran, 6000-2000 B.C., *Chronologies in Old World Archaeology* (R.A. Ehrich). Chicago: 215-56.

Erskine, K.D. (1908). *Rajputana Gazetteers Vol. II. The Mewar Residency*. Ajmer.

_______ (1909). *Rajputana Gazetteers III-A The Western Rajputana States Residency and The Bikaner Agency*. Allahabad.

Fabris, G.S. and Treolar, F.E. (1980). X-ray Fluorescence and Atomic Absorption Analysis of Sarawak Gold Artifacts. *Archaeometry* 22: 93-8.

Faccenna, D. (1964-5). Preliminary Report on the 1963 Excavations Campaign at Barama I (Swat—Pakistan). *EW* 15: 7-23.

Fairservis,W.A. Jr. (1956). Excavations in the Quetta Valley, West Pakistan. *Anthropological Papers of the American Museum of National History* 45. New York.

_______ (1959). Archaeological Surveys in the Zhob and Loralai Districts, West Pakistan. *Anthropological Papers of the American Museum of National History* 47. New York.

_______ (1971). *The Roots of Ancient India*. London.

_______ (1982). Allahdino: A Small Site of the Harappan Civilization. *Harappan Civilization* (G.L. Possehl). New Delhi: 107-12.

Ferriera, G.P. and Gil, F.B. (1981). Elemental Analysis of Gold Coins by Particle Induced X-ray Emission (PIXE). *Archaeometry* 23: 189-97.

Fisher, F.H. and Hewett, J.P. (1884). *Statistical, Descriptive and Historical Account of the North Western Provinces of India* Vol. XIV, Part I. Benares. Allahabad.

Fitzerland, S.V. & Nelson, A.E. (1911, rpt. 1985). *Central Provinces District Gazetteers* Amravati District. Pune.

Flam, L. (1986). Recent Explorations in Sind: Paleography, Regional Ecology, and Prehistoric Settlement Patterns (ca. 4000-2000 B.C.). *Studies in the Archaeology of India and Pakistan*. Delhi: 65-89.

Forbes, J. (1834). *Oriental Memoirs* 2 Vols. London.

Francfort, H.P. (1984a). The Harappan Settlement of Shortugai. *Frontiers of the Indus Civilization* (B.B. Lal and S.P. Gupta). New Delhi: 301-10.

_______ (1984b). The Early Periods of Shortughai (Harappan) and the Western Bactrian Culture of Dashly. *SAA* (B. Allchin). Cambriege: 170-5.

Francis, W. (1904). *Madras District Gazetteers Bellary*. Madras.

Frifelt, K. (1975). On Prehistoric Settlement and Chronology of the Oman Peninsula. *EW* 25: 320-424.

Gale, N.H. and Gale, Z.S. (1981). Lead and Silver in the Ancient Aegean. *Scientific American* 244: 176-92.

Garrick, H.B.W. (1885). Report of a Tour through Behar, Central India, Peshawar and Yusfzai 1881-82. *ASI*.

Gaur, G.S. (1987). Semthan Excavation: A Step toward Bridging the Gap Between the Neolithic and the Kushan Period in Kashmir. *Archaeology and History* Volume I (B.M. Pande and B.P. Chattopadhyay). Delhi: 327-37.

Gaur, R.C. (1983). *Excavations at Atranjikhera*. Delhi.

(1883-4). *Gazetteer of the Hoshiarpur District*. Punjab.

(1883-4). *Gazetteer of the Jhelam District*. Calcutta.

(1880). *Gazetteer of the Bombay Presidency CI Rewa Kantha, Narukot, Cambay and Surat States*. Bombay.

(1915). *Gazetteer of the Mianwali District*. Lahore.

(1884). *Gazetteer of the Jhang District*. Lahore.

(1895). *Gazetteer of the Rawalpindi District 1893-94*. Lahore.

(1892-3). *Gazetteer of the Ambala District 1892-93*. Punjab.

Ghosh, A. (1948). Taxila (Sirkap). 1994-5. *AI* 4: 66-78.

_____ (1951). Rajgir, 1950. *AI* 7: 66-78.

_____ (1953). Exploration in Bikaner. *EW* IV: 31-4.

Ghosh, N.C. (1986). *Excavations at Satanikota 1977-80. MASI* No. 82. New Delhi.

Gilles, K.J.S. and Urche, D.S. (1983). Spectroscopic Studies of Iron and Carbon in Black Surfaced Wares. *Archaeometry* 25: 29-44.

Gogte, V.D. (1982a). Megalithic Iron Smelting at Naikund (Part-II). Efficiency by Three-Probe Resistivity Survey. *Naikund Excavations* (S.B. Deo and A.P. Jamkhedkar). Bombay: 52-5.

Gogte, V.D., Kanetkar, S.M. and Kshirsagar, A. (1982). Efficiency of Megalithic Iron Smelting by Chemical Analysis. *BDCRI* 41: 68-71.

Gordon, D.H. (1950). The Early Use of Metals in India and Pakistan. *Journal of the Royal Anthropological Institute* 80: 50-78.

Goswami, K.G. (1948). *Excavations at Bangarh*. Calcutta.

Grant, C. (1874, rpt. 1984). *The Gazetteer of the Central Provinces of India*. New Delhi.

Gupta, S.P. (1984). Internal Trade of the Harappans. *Frontiers of the Indus Civilization* (B.B. Lal and S.P. Gupta). New Delhi: 417-24.

Guru, S.D. (1976). *Madhya Pradesh District Gazetteers Raigarh*. Bhopal.

Gururaja Rao, B.K. (1978). The Iron Age in Karnataka. *Archaeology of Karnataka* (D.V. Narasimha Murthy). Mysore: 48-61.

Habib, I. (1982). *An Atlas of the Mughal Empire*. Delhi.

________ (1984). The Geographical Background. *The Cambridge Economic History of India* Vol. I. Hyderabad: 1-13.

Hackett, C.A. (1878). Sketch of the Geology of Ulwur. *Gazetteer of Ulwar* (P.W. Powlett). London: 177-83.

Halim, M.A. (1970-1). Excavations at Sarai Khola Part I. *PA* 7:23-89.

Halim, M.A. (1972). Excavations at Sarai Khola Part II. *PA* 8: 1-112.

Hargreaves, H. (1929). *Excavations in Baluchistan Sampur Mound. Mastung and Sohr Damb. Nal. M.A.S.I. No. 35*. New Delhi.

Harrison, J.V. (1941). Coastal Makran. *The Geographical Journal* XCVII: 1-17.

Hasan, M.U. (1986). Recent Archaeological Finds in Baluchistan and Their Implications. South Asian Archaeology Congress, Delhi. (Cyclostyled).

Hayavadana Rao, C. (1927). *Mysore Gazetteer* Vol. I Descriptive. Bangalore.

Hegde, K.T.M. (1967). Chemical and Spectrometric Studies in the Kshatrapa Silver Coins. *The Journal of the Numismatic Society of India*. XXIX: 63-6.

_______ (1968). Source of the Metal in the Lead Coins of the Kshatrapa Period. *Current Science* 18: 518-20.

_______ (1969). Technical Studies in Copper Artifacts from Ahar. *Excavations at Ahar* (H.D. Sankalia, S.B. Deo and Z.D. Ansari). Poona: 225-8.

_______ (1975). Iron Metallurgy. *Excavations at Dhatva* (R.N. Mehta and S.N. Chowdhary). Baroda: 48-56.

_______ (1985). Scientific Studies in Indian Archaeology. *Recent Advances in Indian Archaeology* (S.B. Deo and K. Paddayya). Poona: 100-9.

Hodder, I. (1982). *The Present Past*. London.

Hodder, I. and Orton, C. (1976). *Spatial Analysis in Archaeology*. Cambridge.

Holdich, T.H. (1901). *The Indian Borderland*. London.

_________ (1910). *The Gates of India*. London.

Holjund, Fl. (1989). Some New Evidence of Harappan Influence in the Arabian Gulf. *SAA* 1985 (K. Frifelt and P. Sorensen). London, Riverdale: 49-54.

Holmes, D.A. (1968). The Recent History of the Indus. *The Geographical Journal* 134: 367-81.

Hoste, De la (1840). Memoranda Resecting the Existence of Copper in the Territory of Luz. *JASB* IX: 30-3.

Hunter, W.W. (1876, rpt. 1973). *A Statistical Account of Bengal* II, IV. Delhi.

______ (1877, rpt. 1976). *A Statistical Account of Bengal* XI, XII, XIV. Delhi. *Indian Archaeology—A Review.*

Jackson, A.V. (1921, rpt. 1955). The Persian Dominions in Northern India down to the Time of Alexander's Invasion. *The Cambridge History of India* Vol. I. Delhi: 285-306.

Jacobson, J. (1975). Static Sites and Peripatetic Peoples: Notes on the Archaeology of Population Mobility in Eastern Malwa. *Pastoralists And Nomads in South Asia* (L.S. Leshnik and G.D. Sontheimer). Weisbaden: 68-91.

Jain, R.S. (1987). Gold Incidence in Rajasthan. *Records of the G.S.I.* 113: 99-103.

Jarrige, J.F. (1974-86). Excavations at Mehrgarh-Nausharo. *PA* 10-22: 63-131.

______ (1981). Economy and Society in the Early Chalcolithic/Bronze Age of Baluchistan: New Perspectives from Recent Excavations at Mehrgarh. *SAA* 1979 (H. Hartel). Berlin: 93-114.

______ (1984). Chronology of the Earlier Periods of the Greater Indus as seen from Mehrgarh, Pakistan. *SAA* 1981 (B. Allchin). Berlin: 21-8.

______ (1985a). Foreword. *Pre-Harappan Cultures of India and the Borderlands* (S. Asthana). New Delhi: v-viii.

______ (1985b). Continuity and Change in the North Kachi Plain (Baluchistan, Pakistan) at the Beginning of the second millennium B.C. *SAA* 1983 (J. Schotsmans and M. Taddei). Naples: 35-68.

Jarrige, J.F. and Enault, J.F.(1973). Recent Excavations (French) in Pakistan. (D.P. Agrawal and A. Ghosh). *Radiocarbon and Indian Archaeology.* Bombay: 163-72.

Jarrige, J.F. and Lechevallier, M. (1979). Excavations at Mehrgarh, Baluchistan: Their Significance in the Prehistorical Context of the Indo-Pakistan Borderlands. *SAA* 1977(M. Taddei). Naples: 463-535.

Jauhari, A.S. and Verma, M.S. (1987). Punjab Plain. *India: A Regional Geography* (R.L. Singh). Varanasi: 83-123.

Joshi, E.B. (1968). *Uttar Pradesh Gazetteer Allahabad.* Allahabad.

Joshi, J.P. (1966). Exploration in Northern Kutch. *JOIB* LXXI: 62-9.

_______ (1972).Exploration in Northern Kutch and Excavation at Surkotada. *JOIB* 22: 98-144.

Joshi, J.P. and Bala, M. (1982). Manda: A Harappan Site in Jammu and Kashmir. *Harappan Civilization* (G.L. Possehl). New Delhi: 185-96.

Joshi, J.P., Bala, M. and Ram, J. (1984). The Indus Civilization: A Reconsideration on the Basis of Distribution Maps. *Frontiers of the Indus Civilization* (B.B. Lal and S.P. Gupta). Delhi: 511-30.

Kamili, M.H. (1967). *Census of India.* 1961 Vol. VI—Part IX Census Atlas of Jammu and Kashmir. Srinagar.

Kangle, R.P. (1963). *The Kautilya Arthasastra.* Bombay.

Kareem, C.K. (1976). *Kerala District Gazetteers Palghat.* Trivandrum.

Karunakaran, C. (1974a). *Geology and Mineral Resources of the States of India Part VII. Karnataka, Goa, Daman and Diu.* G.S.I. Miscellaneous Publication No. 30. Delhi.

______ (1974b). *Geology and Mineral Resources of the States of India Part VI—Tamil Nadu and Pondicherry* G.S.I. Miscellaneous Publication No. 30. Delhi.

______ (1975). *Geology and Mineral Resources of the States of India Part VIII— Andhra Pradesh.* G.S.I. Miscellaneous Publication No. 30. Delhi.

Khan, A.R. (1981). 'Kot Kori'—A Site on Indus Civilization in the Lower Sind. *Indus Civilization—New Perspectives.* Islamabad: 43-6.

Khan, A.W. (1963). *A Monograph on Yeleswaram Excavations.* Hyderabad.

Khan, F.A. (1965). Excavations at Kot Diji. *PA* 2: 13-85.

Khan, F. Knox, J.R. and Thomas, K.D. (1986). Sheri Khan Tarakai: A New Site in the North West Frontier Province of Pakistan. *Journal of Central Asia* IX: 13-33.

_______ (1989). New Perspectives on Early Settlement in Bannu District, Pakistan. *SAA* 1985 (K. Frifelt and P. Sorenson). London: Riverdale: 281-91.

Khare, M.D. (1980-1). Neo-Chalcolithic Remains at Rangai (Vidisha). *Puratattva* 12: 55-7.

_______ (1982). Chalcolithic Besnagar (Vidisha). *Indian Archaeology New Perspectives* (R.K. Sharma). Delhi: 220-3.

Khera, B. (1982). A Newly Discovered Chalcolithic Site in the Middle Narmada Valley. *Indian Archaeology New Perspectives* (Sharma, R.K.). Delhi: 268-9.

Khunte, B.G. (1974a). *Maharashtra State Gazetteers Dhulia District*. Bombay.

_______ (1974b). *Maharashtra State Gazetteers Wardha District*. Bombay.

_______ (1975). *Maharashtra District Gazetteers Nasik District*. Bombay.

_______ (1976). *Maharashtra State Gazetteers Ahmadnagar District*. Bombay.

Krishna Sastry, V.V. (1983). *The Proto and Early Historical Cultures of Andhra Pradesh*. Hyderabad.

Krishnamurthy, C. (1978). Recent Discoveries in South Karnataka. *Archaeology of Karnataka* (A.V. Narasimha Murthy). Mysore: 80-9.

Krishnan, M.S. (1951). Mineral Resources of Madras. *M.G.S.I.* Vol. 80. Calcutta.

Krishnan, V.S. (1965). *Madhya Pradesh District Gazetteers Gwalior*. Bhopal.

_______ (1967). *Madhya Pradesh District Gazetteers Sagar*. Bhopal.

Krishnaswami, V.D. (1949). Megalithic Types of South India. *AI* 5: 35-45.

Kumar, G. and Dhaundiyal, J.N. (1977). Lead-Silver Mineralisation in Birgana Area, Eastern Nayar Valley, Garhwal District, Uttar Pradesh. *IM* 31: 5-8.

Lahiri, N. (1990). Harappa as a Centre of Trade and Trade Routes: A Case Study of the Resource-Use, Resource-Access and Lines of Communication in the Indus Civilization. *Indian Economic and Social History Review* 27, 4: 405-44.

Lal, B.B. (1951). Further Copper Hoards from the Gangetic Basin and a Review of the Problem. *AI* 7: 20-39.

______ (1954-5). Excavations at Hastinapura and Other Explorations in the Upper Ganga and Sutlej Basins 1950-52. *AI* 10 & 11: 5-151.

Lal, B.B. (1949). Sisupalgarh 1948: An Early Historical Fort in Eastern India. *AI* 5: 62-105.

Lal, B.B. (1972). The Copper Hoard Culture of the Ganga Valley. *Antiquity* VLVI: 232-87.

Lal, B.B. and Dikshit, K.N. (1978-9). Sringaverapura: A Key—Site for the Protohistory and Early History of the Central Ganga Valley. *Puratattva* 10: 1-7.

Lal, J.K., Srivastava, Saxena, A. & Lari, T.A. (1985). A Note on Palaeo— and Neo Placer Gold Occurrences in Siwalik Belt of Western Himalaya. *IM* 39: 22-4.

Lal, M. (1980-1). Copper Hoard Culture of India: A Reassessment. *Puratattva* 12: 65-77.

______ (1984a). Summary of Four Seasons of Exploration in Kanpur District, Uttar Pradesh. *ME* VIII: 61-80.

______ (1984b). *Settlement History and Rise of Civilization in the Ganga-Yamuna Doab*. Delhi.

______ (1986). Iron tools, Forest Clearance and Urbanization in the Gangetic Plains. *ME* X: 83-90.

Lamberg-Karlovsky, C.C. and Tosi, M. (1973). Shahr-I Sokhta and Tepe Yahya: Tracks on the Earliest History of the Iranian Plateau. *EW* 23: 21-57.

Lambrick, H.T. (1964). *Sind A General Introduction*. Hyderabad.

La Touche, T.H.D. (1918). *A Bibliography of Indian Geology and Physical Geography with an Annotated Index of Minerals of Economic Value*. 2 Vols. Calcutta.

Law, B.C. (1954). *Historical Geography of Ancient India*. France.

Lawrence, W.A. (1895). *The Valley of Kashmir*. London.

La Fanu,H. (1883). *A Manual of the Salem District in the Presidency of Madras*. Madras.

Leshnik, L.S. (1974). *South Indian Megalithic Burials*. Weisbaden.

Losch, A. (1954). *The Economics of Location*. Yale.

Mackay, E.J.H. (1938). *Further Excavations at Mohenjdaro* 2 Vols. Delhi.

______ (1943). *Chanhu-daro Excavations*. Delhi.

Madhava Rao, P.S. (1961). *Maharashtra State Gazetteers Nagpur District*. Bombay.

Mahapatra, R.P. (1986). *Archaeology in Orissa* 2 Vols. Delhi.

Majumdar, N.G. (1934). *Explorations in Sind*. *MASI* No. 48. New Delhi.

Malcolm, J. (1823, rpt. 1970). *A Memoir of Central India Including Malwa, and Adjoining Provinces*. 2 Vols. Delhi.

Malik, S.C (1972-3). The Role of Theory in the Study of Archaeology in India. *Puratattva* 6: 1-11.

Mallett, F.R. (1887). *A Manual of the Geology of India. Part IV. Mineralogy.* Calcutta.

Mallett, F.R. (1951). *Taxila*. 3 Vols. Cambridge.

Marshall, J.H., Vogel, J. Ph. (1904, rpt. 1970). Excavations at Charsada in The Frontier Province. *ARASI* 1902-3. Varanasi: 141-84.

Martin, M. (1838a, rpt. 1976). *The History, Antiquities, Topography and Statistics of Eastern India* Vol. II Bhagalpur, Gorakhpur. Delhi.

_______ (1838b, rpt. 1976). *The History, Antiquities Topography and Statistics of Eastern India* vol. I Bihar (Patna City) and Shahabad. Delhi.

Massie, J.W. (1840, rpt. 1985). *Continental India* 2 vols. Delhi.

Masson, C. (1842, rpt. 1975). *Narrative of Various Journeys in Baluchistan, Afghanistan and the Punjab... from 1826 to 1858.* Austria.

_______ (1843). *Narrative of a Journey to Kalat... And a Memoir on Eastern Baluchistan.* London.

M'Crindle, J.W. (MDCCCDVI). *The Invasion of India by Alexander the Great.* Westminster.

Mehta, R.N. (1968). *Excavation at Nagara.* Baroda.

_______ (1955). *Excavations at Timbarva.* Baroda.

Mehta, R.N. and Chowdhary, S.N. (1975). *Excavations at Dhatva.* Baroda.

Mehta, R.N., Chowdhary, S.N., Hegde, K.T.M. and Shah, D.R. (1971). *Excavations at Jokha.* Baroda.

Mehta, R.N., Momin, K.N. and Shah, D.R. (1980). *Excavations at Kanewal.* Baroda.

Minchin, C.F. (1907a). *Baluchistan District Gazetteer Series II Loralai District.* Allahabad.

_______ (1907b). *Baluchistan District Gazetteer Series. Zhob District.* Bombay.

_______ (1907c). *Baluchistan District Gazetteer Series VI VI-A and VI-B. Sarawan, Kachhi and Jhalawan.* Bombay.

_______ (1907d). *Baluchistan District Gazetteers Series VIII Las Bela.* Allahabad.

Mishra, A.K. (2003). *Ethnoarchaeometallurgy of Iran in India.* Delhi.

Moti Chandra (1977). *Trade and Trade Routes in Ancient India*. New Delhi.

Mughal, M.R. (1967a). Excavations at Tulamba, West Pakistan. *PA* 4: 12-158.

______ (1967b). Exploration in West Pakistan. *PA* 4: 1-3.

______ (1971). *The Early Harappan Period in the Greater Indus Valley and Northern Baluchistan* (*C. 3000-2400 B.C.*) University Microfilms. Michigan.

______ (1972a). Exploration inNorthern Balucistan.*PA* 8:137-50.

______ (1972b). Explorations in Northern Punjab, Campbellpur and Rawalpindi Districts. *PA* 8: 131-2.

______ (1972c). Excavations at Jalilpur. *PA* 8:117-24.

______ (1981). New Archaeological Evidence from Bahawalpur. *Indus Civilization—New Perspectives* (A.H.Dani). Islamabad: 33-41.

Mughal, M.R. (1982). Recent Archaeological Research in the Cholistan Desert. *Harappan Civilization* (G.I. Possehl). New Delhi: 85-96.

Muhammad Din, M. (1908). *Punjab State Gazetteers XXXVI A. Bahawalpur State*. Lahore.

Mukherjee, M. (1978). *Metalcraftsmen of India*. Anthropological Survey of India. Memoir No. 44. Calcutta.

Nagaraja Rao, M.S. (1962-3). Recent Explorations in the Tungabhadra Basin.*BDCRI* 23: 54-77.

______ (1971). *Protohistoric Cultures of the Tungabhadra Valley* (A Report on the HALLUR Excavations). Dharwar.

Nanavati, J.M., Mehta, R.N. and Chowdhary, S.N. (1971). *Somnath—1956*. Baroda.

Nanda, R.and Chakrabarti, D.K. (1981). Lead in Early India: A Study of the Geological, Archaeological and Literary Data. *IHR VIII* : 1-18.

Narain,A.K. and Roy T.N. (1976). *Excavations at Rajghat*. Part I. Varanasi.

_____ (1977). *Excavations at Rajghat*. Part II. Varanasi.

Narasimhaiah, B. (1978). Recent Discoveries in the Archaeology of North Karnataka. *Archaeology of Karnataka* (A.V. Narasimha Murthy). Mysore: 90-101.

_____ (1980). *Neolithic and Megalithic Cultures in Tamil Nadu*. Delhi.

Nautiyal, K.P. (1969). *The Archaeology of Kumaon* (including Dehradun). Varanasi.

_____ (1989). *Protohistoric India*. Delhi.

Nautiyal, K.P., Khanduri, K.P., Nautiyal, V. and Rajput, D.L. (1978-9). Thapli: A Painted Grey Ware Site in Garhwal. *Puratattva* 19: 94-5.

Nautiyal, K.P., Khanduri, B.M., Naitiyal, V., Rajput, D.L. and Bhatt, R.K. (1986-7). Painted Grey Ware Culture in Garhwal, Himalaya: New Evidence and Interpretation. *Puratattva* 17: 11-14.

Nautiyal, V., Agrawal, D.P. and Krishnamurthy, R.V. (1981). Some New Analyses on the Protohistoric Copper Artifacts. *ME* 5: 48-51.

Negi, R.S. (1982). Population Structure of a Pastoral Nomadic Community of the Himalayas. *Nomads in India* (P.K. Misra and K.C. Malhotra). Calcutta: 42-51.

Nevill, H.L. (1910, rpt. 1977). *Campaigns on the North-West Frontier*. Pakistan.

Nevill, H.R. (1904). *Nainital: A Gazetteer... Agra and Oudh*. Allahabad.

_____ (1905). *Agra A Gazetteer... Agra and Oudh*. Allahabad.

_____ (1907). *Budaun A Gazetteer... of Agra and Oudh*. Allahabad.

_____ (1909). *Hardoi: A Gazetteer... Agra and Oudh*. Lucknow.

_____ (1920). *Cawnpore A Gazetteer... Agra and Oudh*. Allahabad.

_____ (1928). *Bijnor A Gazetteer... Agra and Oudh*. Allahabad.

Nigam, M.L. (1971). *Report on the Excavation of Two Megalithic Burials at Hashmatpet. Hyderabad*. Hyderabad.

Oldham, R.D. (1893). *A Manual of the Geology of India*. Stratigraphical and Structural Geology. Calcutta.

O'Malley, L.S.S. (1907a). *Bihar District Gazetteer Series Champaran*. Calcutta.

______ (1907b). *Bihar District Gazetteer Series Patna*. Calcutta.

______ (1910a). *Bihar District Gazetteer Series Singhbhum. Saraikela and Kharsawan*. Calcutta.

Paddayya, K. (1971). Radiocarbon Dates and South Indian Neolithic Cultures. *Antiquity* 45: 134-38.

______ (1973). *Investigations into the Neolithic Culture of the Shorapur Doab. South India*. London.

______ (1981-3). Ecological Archaeology and the Ecology of Archaeology: The Archaeologist's Viewpoint. *Puratattva* 13 & 14: 130-50.

______ (1985). Theoretical Archaeology—A Review. *Recent Advances in Indian Archaeology* (S.B. Deo and K. Paddayya). Poona: 6-22.

Palande, M.R. (1954). *Gazetteer of the Bombay State Poona District Series—Vol. XX*. Bombay.

______ (1959). *Gazetteer of Bombay State Dharwar District*. Bombay.

______ (1961). *Gujarat State Gazetteer Broach District*. Ahmedabad.

Pant, C.C. and Mishra, D.K. (1983). Mineral Resources of Kumaon Himalaya. *The Himalaya: Nature, Man and Culture* (O.P. Singh). New Delhi: 174-83.

Pascoe, E. (1931). Minerals and Metals. *Mohenjodaro and The Indus Civilization* (J. Marshall) Vol. II. Delhi: 674-91.

Patel, G.D. (1971). *Gujarat State Gazetteers Kutch District*. Ahmedabad.

______ (1972a). *Gujarat State Gazetteers Pachmahals District*. Ahmedabad.

______ (1972b). *Gujarat State Gazetteers Amreli District*. Ahmedabad.

______ (1977). *Gujarat Gazetteers Surendranagar District*. Ahmedabad.

Pires Ferriera, J.W. and Flannery, K.V. (1976). Ethnographic Models Formative Exchange. *The Early Mesoamerican Village*. New York: 286-92.

Plog, F. (1977). Modelling Economic Exchange. *Exchange Systems in Prehistory* (T.K. Earle and J.E. Ericson). New York: 127-40.

Pollard, A.M., Russel, G.D. and Baud, D.C. (1981). The Analytical Investigations of Early Bronze Age Jet and Jet-like Material from the Devizes Museum. *Archaeometry* 23: 131-67.

Possehl, G.L. (1980). *Indus Civilization in Saurashtra*. Delhi.

_______ (1984). A Note on Harappan Settlement Patterns in the Punjab. *Studies in the Archaeology and Palaeoanthropology of South Asia* (K.A.R. Kennedy and G.L. Possehl). New Delhi: 83-7.

_______ (1986). *Kulli—An Exploration of Ancient Civilization in Asia*. Durham.

_______ (1988). Radiocarbon Dates from South Asia. *ME* XII: 169-96.

Powlett, P.W. (1878). *Gazetteer of Ulwar*. London.

Pracchia, S., Tosi, M. and Vidale, M. (1985). On the Type, Distribution and Extent of Craft Industries at Mohenjo-daro. *SAA* 1983 (J. Schotsmans and M. Taddei). Naples: 207-47.

Prasad, N. and Singh, O. (1987). Andhra Plateau. *India A Regional Geography*. Varanasi: 821-50.

Prinsep, V.G. (1876). *Imperial India, an Artist's Journal*. London.

Prufer, O. (1952). *Nalagarh 1951*. Interim Report on the Excavation Carried Out at Dher Majra. Jamia Millia Islamia Historical Research Foundation.

_____ (1932). *Punjab District Gazetteers Vol. XXIX Attock District*. A 1930. Lahore.

Raghu Nandan, K.R., Dhruva Rao, B.K. and Singhal, M.L. (1981). *Exploration for Copper, Lead and Zinc Ores in India*. Bulletins of the G.S.I. Series A—*Economic Geology* No. 47. Calcutta.

Raina, A.N. (1977). *Geography of Jammu and Kashmir*. New Delhi.

Rajagopal, M.V. (1974). *Andhra Pradesh District Gazetteers Kurnool*. Hyderabad.

_______ (1976a). *Andhra Pradesh District Gazetteers Mahbubnagar*. Hyderabad.

______ (1976b). *Andhra Pradesh District Gazetteers Medak*. Hyderabad.

______ (1977a). *Andhra Pradesh District Gazetteers Guntur*. Hyderabad.

______ (1977b). *Andhra Pradesh District Gazetteers Krishna*. Hyderabad.

______ (1978). *Andhra Pradesh District Gazetteers Nalgonda*. Hyderabad.

Rajyagor, S.B. (1974). *Gujarat State Gazetteers Sabarkantha District*. Ahmedabad.

______ (1975). *Gujarat State Gazetteers Junagadh District*. Ahmedabad.

______ (1977). *Gujarat State Gazetteers Kheda District*. Ahmedabad.

Rajyagor, S.B. and Tripathy, S. (1979). *Gujarat State Gazetteers Vadodara District*. Ahmedabad.

Ramachandran, K.S. (1974). Cairn Burials of Baluchistan, Makran and Southeastern Iran, Their Character and Chronology. *Puratattva* 7: 82-7.

______ (1980. *Archaeology of South India Tamil Nadu*. Delhi.

Ramesh, A. and Gopalakrishnan, K.S. (1987). Tamil Nadu Uplands and South Sahyadri. *India A Regional Geography* (R.L. Singh). Varanasi: 851-78.

Rao, M.H. and Nagaraju, S. (1974). *Excavations at Hemmige*. Mysore.

Rao, P.R. and Goswai, N.G. (ed.) (2001). *Metallurgy in India—A Retrospective.*

Rao, S.R. (1962 & 1963). Excavations at Rangpur and Other Explorations in Gujarat. *AI* 18 & 19: 5-207.

Rao, S.R. (1979-85). *Lothal A Harappan Port Town 2 Vols. MASI* No. 78. New Delhi.

Ratnagar, S. (1981). *Encounters, the Westerly Trade of the Harappan Civilization*. Delhi.

________ (1982). The Location of Harappa. *Harappan Civilization* (G.L. Possehl). New Delhi: 261-4.

Rawlinson, H.G. (1908 to 1913). Foreign Influences in the Civilization of Ancient India 900 B.C. – 400 A.D. *JARS* XXXIII: 27-238.

________ (1916). *Intercourse Between India and the Western World*. Cambridge.

Ray, S.C. (1957). *Early History and Culture of Kashmir*. Calcutta.

Raychaudhuri, H.C. (1960). A Note on the New Incidences of Gold, Tungsten, Niobium-Titanium and Other Rare Mineralization in the Sakori Basin, Nagpur and Bhandara Districts, Maharashtra. *IM* 35: 39-41.

Rennell, J. (1793, rpt. 1976). *Memoir of a Map of Hindoostan or the Moghal Empire*. Calcutta.

Rice, L. (1876). *Mysore and Coorg* Vol. II. Bangalore.

____ (1897). *Mysore A Gazetteer* Compiled for Government, Vol. I. West minster.

Richards, F.J. (1924). Anthropological Geography. *Man in India* IV: 14-40.

_____ (1933). Geographical Factors in Indian Archaeology. *Indian Antiquary* 52: 235-43.

Rissman, P. (1985). The Oriyo Test Excavation and the End of the Harappan Tradition in Gujarat. *SAA* 1983 (J. Schotsmans and M. Taddei). Naples: 345-55.

Rowlands, M.J. (1971). The Archaeological Interpretation of Prehistoric Metal Working. *WA* 3: 210-24.

Roy, B.C. (1959). *The Economic Geology and Mineral Resources of Rajasthan and Ajmer. M.G.S.I.* No. 86. Delhi.

Roy Chaudhury, P.C. (1960). *Bihar District Gazetteers Saran.* Patna.

Roy, Choudhury, M.K. (1974). *Geology and Mineral Resources of the States of India Part II – Maharashtra.* Delhi.

Russell, R.V. (1908). *Central Provinces District Gazetteers. Nimar District.* Allahabad.

Sahi, M.D.N. (1979). Iron at Ahar. *Essays in Indian Protohistory* (D.P. Agrawal and D.K. Chakrabarti). Delhi: 365-8.

Sahni, D.R. (n.d.). Archaeological Remains and Excavations at Bairat.

Sali, S.A. (1986). *Daimabad, MASI* No. 83. New Delhi.

Sanaullah, M. (1940, rpt. 1974). The Sources, Composition and Technique of Copper and Its Alloys. *Excavations at Harappa* (M.S. Vats). Varanasi: 378-91.

Sankalia, H.D. (1977). The Lifeways and Settlement Pattern at Inamgaon, a Chalcolithic Site, District, Poona, Maharashtra.*ME* I: 44-5.

Sankalia, H.D. (1977). *New Archaeology and Its Scope and Application in India.* Lucknow.

Sankalia, H.D. and Deo, S.B. (1955). *Report on the Excavations at Nasik and Jarwe.* Poona.

Sankalia, H.D., Ansari, Z.D. and Dhavalikar, M.K. (1975). An Early Farmers' Village in Central India. *Expedition* 17: 2-11.

Sankalia, H.D., Deo, S.B. and Ansari, Z.D. (1969). *Excavations at Ahar.* Poona.

______ (1971). *Chalcolithic Navdatoli.* Poona. Baroda.

______ (1960). *From History to Pre-History at Nevasa (1954-56).*Poona.

Sankalia, H.D., Subbarao, B. and Deo, S.B. (1958). *The Excavations at Maheshwar and Navdatoli* 1952-53. Baroda.

Sankaranarayana, B.S. (1967). *Andhra Pradesh District Gazetteers Cuddapah.* Hyderabad.

______ (1970). *Andhra Pradesh District Gazetteers Anantapur.* Hyderabad.

Santoni, M. (1984). Sibri and the South Cemetery of Mehrgarh: Third Millennium Connections between the Northern Kachi Plain (Pakistan) and Central Asia. *SAA* 1981 (B. Allchin). Cambridge: 52-60.

Sastri, T.V.G. (1961). Cultural Heritage of Nagarjunakonda. *Journal of the Oriental Institute* XI: 9-18.

Satyanarayan Rao, P. (1972). The Geology and Copper Occurrences of Ganikalavaveklurti Area.... *Bulletins of the G.S.I. No. 36.* Delhi.

Scott, D.A. (1983). A Note on the Dissolution of Ancient Gold Alloys. *Archaeometry* 25: 214-16.

Scott, F.H. (1853). *Routes in the Peninsula of India.* Madras.

Seely, N.J. and Turner, P.J. (1984). Metallurgical Investigations of Three Early Indian Coinages: Implications for Metal Trading and Dynastic Chronology. *SAA* 1981. Cambridge: 331-3.

Sehgal, KK. (1972a). *Rajasthan District Gazetteers Bikaner.* Jaipur.

______ (1972b). *Rajasthan District Gazetteers Ganganagar.* Jaipur.

Senapati N. and Kuanr, D.C. (1975). *Orissa District Gazetteers Sundergarh.* Bhubaneshwar.

______ (1980). *Orissa District Gazetteers Kalahandi.* Bhubaneshwar.

Serge, C. (1984). Oman Peninsula and Its Relations Eastward during Third Millennium. *Frontiers of the Indus Civilization* (B.B. Lal and S.P. Gupta). New Delhi: 371-94.

Seshadri, M. (1971a). *Report on the Excavations at T. Ņarasipur.* Bangalore.

______ (1971b). Chalcolithic Culture at T. Narasipur. *Studies in Indian History and Culture* (S. Ritti and B.R. Gopal). Dharwar: 1-4.

Shaffer, J.G. (1978). *Prehistoric Baluchistan.* Delhi.

______ (1979). The Indus Civilization: New Evidence from Pakistan. *Essays in Indian Protohistory* (D.P. Agrawal and D.K. Chakrabarti). Delhi: 17-29.

Shaffer, J.G. (1980). Harappan Commerce: An Alternative Perspective. To be published in *Current Anthropology in Pakistan* (L. Flam and S. Papsner). Ithaca.

______ (1981). The Protohistoric Period in the Eastern Punjab: A Preliminary Assessment. *Indus Civilization—New Perspectives* (A.H. Dani). Islamabad: 65-102.

______ (1986). Cultural Development in the Eastern Punjab. *Studies in the Archaeology of India and Pakistan*. (J. Jacobson). New Delhi: 195-235.

Sharif, M. (1969). Excavation at Bhir Mound Taxila. *PA* 6: 7-99.

Sharma, G.B. and Kumar, M. (1980-1). Excavations at Rohita. *Puratattva* 12:125-7.

Sharma, G.N. (1985). *Rajasthan Studies*. Agra.

Sharma, G.R. (1968). *The Excavations at Kausambi 1957-59*. Allahabad.

______ (1969). *Excavations at Kausambi 1949-50*. *MASI* No. 74. Delhi.

Sharma, R.K. and Ali, R. (1980). *Archaeology of Bhopal Region*. Delhi.

Sharma, Y.D. (1953). Exploration of Historical Sites. *AI* 9: 116-69.

______ (1964). *Protohistoric Remains, Archaeological Remains, Monuments and Museums*. New Delhi.

______ (1981). Presidential Address (Ancient Section). *Punjab History Conferences... 1980 Proceedings*. Patiala. 23-32.

______ (1982). Harappan Complex on the Sutlej (India). *Harappan Civilization* (G.L. Possehl). New Delhi: 141-66.

Sharma, Y.D. and Kumar, M. (1980). Excavation at Ghuram. *Punjab History Conference.... 1979 Proceedings*. Patiala: 32-5.

Shekhar, N.C. and Mukul, S.S. (1972). Occurrence of Base Metal Ores at Banejnes, Gir Forest, Junagadh District, Gujarat. *G.S.I. Miscellaneous Publication No. 16 Base Metals (Part II)*. Delhi: 553-61.

Sherwani, H.K. and Joshi, P.M. (1973). *History of Medieval Deccan* Vol. I. Hyderabad.

Shinde, V. (1985). Kaothe—A Late Harappan Settlement in the Central Tapti Basin. *BDCRI* 44: 173-7.

Shrivastava, K.K. and Saha, H.L. (1975). A Recent Find of Deccan Traps near Sala, Jalaun District, Uttar Pradesh, *Records of the G.S.I.* 106: 102-9.

Shrivastava, P.N. (1968). *Madhya Pradesh District Gazetteers Jabalpur*. Bhopal.

_____ (1969). *Gazetteer of India Madhya Pradesh East Nimar*. Bhopal.

_______ (1970). *Madhya Pradesh District Gazetteers West Nimar*. Bhopal.

Shrivastava, P.N. and Sharma, R. (1979). *Madhya Pradesh District Gazetteers Vidisha*. Bhopal.

Shrivastava, R.N. and Kapoor, A.K. (1981). Argentiferous Lead Mineralisation in Birgna Village, Pauri-Garhwal District. Uttar Pradesh. *IM* 35:8-12.

Singh, H.N. (1982). *History and Archaeology of Black And Red Ware (Chalcolithic Period)*. Delhi.

Singh, J. and Thapan, D. (1987). Chotanagpur Region. *India A Regional Geography* (R.L. Singh). New Delhi: 3-30.

Singh, O.P. (1983). The U.P. Himalaya: An Overview.*The Himalaya: Nature, Man and Culture* (O.P. Singh). New Delhi: 3-30.

Singh, P. and Lal, M. (1985). Narhan 1983-85: A Preliminary Report. *Bharati* 3:11.

Singh, P. and Lal. M. and Singh, A.K. (1983-5). Excavations at Narhan 1983-85. *Puratattva* 15: 117-20.

Singh, P., Lal, M. and Singh, A.K. (1986-7). The 1985 excavation at Narhan. *Bulletin of the Indo-Pacific Prehistory Association* 7: 27-9.

Singh, R.L. (1987). *India A Regional Geography*. Varanasi.

Singh, R.L., Singh, K.N., Vishwanath, M.S. and Singh, S.N. (1987). Karnataka Plateau. *India A Regional Geography* (R.L. Singh). Varanasi: 791-820.

Singh, U., Verma, R.V., Singh, S.P. and Hira Lal (1987). Upper Ganga Plain. *India A Regional Geography* (R.L. Singh). Varanasi: 124-82.

Sinh, R. (1936). *Mahwa in Transition*. Bombay.

Sinha, K.K. (1967). *Excavations at Sravasti-1959*. Varanasi.

Sivadas, K.M., Sashi Kumar, K.T., Subha Rao, N., Setti, D.N., Rajvekar, S.T., Sharma, R.K., Gopalakrishnan, K.P., Sagar, A.K. ,Ziauddin, M. Murthy, Y.G.K. & Narayanaswamy (1985). Lead and Copper Deposits of the Agnigundala Mineralized Belt – *M.G.S.I. 118*. Calcutta.

Smyth, J.W. (1919a). *Gazetteer of the Province of Sind II Sukkur District*. Bombay.

_____ (1919b). *Gazetteer of the Province of Sind Larkana District*. Bombay.

_____ (1920a). *Gazetteer of the Province of Sind Nawabshah District*. Bombay.

Soundara Rajan, K.V. (1969). *Seminar Papers on the Problem of Megaliths in India*. Varanasi.

_______ (1980). Chrono-Cultural Reconstruction of Kaveripattinam (Tamil Nadu). South Asian Archaeological Congress. Delhi. (Cyclostyled).

Spate, O.H.K. (1964). *India and Pakistan*. London. New York.

Sreedhara Menon, A. (1962a). *Kerala District Gazetteers Kozhikode*. Trivandrum.

_____ (1962b). *Kerala District Gazetteers Trichur*. Trivendrum.

_____ (1972). *Kerala District Gazetteers Cannanore*. Trivandrum.

Srinivasan, K.R. (1946). The Megalithic Burials and Urn-Fields of South India in the Light of Tamil Literature and Tradition. *AI* 2: 9-16.

Srinivasan, K.R. and Banerjee, N.R. (1953). Survey of South Indian Megaliths. *AI* 9: 103-15.

Sri Sathyan, B.N. (1965). *Mysore State Gazetteer Coorg District*. Bangalore.

_____ (1966a). *Mysore State Gazetteer Gulbarga District*. Bangalore.

_____ (1966b). *Mysore State Gazetteer Bijapur District*. Bangalore.

_____ (1967). *Mysore State Gazetteer Chitradurga District*. Bangalore.

Stacul, G. (1969b). Discovery of Protohistoric Cemetries in the Chitral Valley (West Pakistan). *EW* 19: 92-9.

_____ (1970a). The Grey Pottery in the Swat Valley and the Indo-Iranian Connections (1300-300 B.C.). *EW* 20: 92-102.

_____ (1970b). An Archaeological Survey near Kalam (Swat Kohistan). *EW* 20: 87-91.

_____ (1973). Ochre-Coloured and Grey-Burnished Wares in North-West Indo-Pakistan (c. 1800-1300 B.C.). *EW* 23:79-88.

_____ (1975). The Fractional Burial Customs in the Swat Valley and Some Connected Problems. *EW* 25: 323-5.

_____ (1977). Dwelling and Storage Pits at Loebanr III (Swat, Pakistan) 1976 Excavation Report. *EW* 27: 227-53.

_____ (1978). Excavation at Bir-kot-ghundai (Swat, Pakistan). *EW* 28: 137-50.

_____ (1979a). Early Iron Age in the North West of Sub-continent. *Essays in Indian Protohistory* (D.P. Agarwal and D.K. Chakrabarti). Delhi: 341-5.

Stacul, G. (1979b). The Black Burnished Ware Period in the Swat Valley (c. 1700-1500 B.C.). *SAA* 1977 (M. Taddei) Naples: 661-73.

_____ (1979c). The Sequence of the Proto-historical Periods at Aligrama (Swat, Pakistan). *SAA* 1975: (M. Taddei). Naples: 88-90.

_____ (1980a). Bir-kot-ghundai (Swat, Pakistan) 1978 Excavation Report. *EW* 30: 55-65.

_____ (1980b). Loebanr III (Swat, Pakistan), 1979 Excavation Report. *EW* 30: 67-76.

_____ (1981). Painted Pottery from the Swat Valley, Pakistan. *SAA* 1979 (H. Hartel). Berlin: 306-11.

_____ (1984a). Cultural Change in the Swat Valley and Beyond, c. 3000-1400 B.C. *SAA* 1981 (B. Allchin). Cambridge: 205-12.

_____ (1984b). Harappan Post-Urban Evidence in the Swat Valley. *Frontiers of the Indus Civilization* (B.B. Lal and S.P. Gupta). Delhi:271-5.

_____ (1985). A Harappan Post-Urban Outpost in the Swat Valley. *SAA* 1983 (J. Schotsman and M. Taddei). Naples: 357-67.

Stacul, G. and Tusa, S. (1975). Report on the Excavations at Aligrama (Swat, Pakistan), 1966, 1972. *EW* 25: 291-322.

Stacul, G. and Tusa, S. (1977). Report on the Excavations at Aligrama (Swat, Pakistan) 1974, *EW* 27: 151-206.

Steel, C.D. (1884). *Statistical Descriptive, and Historical Account of the North Western Provinces of India*, vol. VIII. Allahabad.

Subbaiah, K.V. & Rao, S. (1984). Chemical Composition of Copper Coins Belonging to the Barid Shahi, Nizam Shahi, Adil Shahi and Qutub Shahi Dynasties. *Journal of Archaeological Chemistry*. 12: 7-30.

Subbaraman, S. (1974-5). Some Modern Analytical Techniques in the Examination of Cultural Property. *Studies in Museology* X: 69-78.

Sundara, A. (1973). Chronological Order of Megalithic Tomb-types in Karnataka. *Radiocarbon and Indian Archaeology* (D.P. Agarwal and A. Ghosh). Bombay: 239-51.

_____ (1975). *The Early Chamber Tombs of South India*. Delhi.

_______ (1979). Typology of Megaliths in South India. *Essays in Indian Protohistory* (D.P. Agrawal and D.K. Chakrabarti). Delhi: 331-40.

Thapar, B.K. (1952). Porkalam 1948 Excavation of a Megalithic Urn-burial. *AI* 13: 4-142.

_______ (1957). Maski 1954: A Chalcolithic Site of the Southern Deccan. *AI* 13: 4-142.

_______ (1964 & 65). Prakash 1955: A Chalcolithic Site in the Tapti Valley. *AI* 20 & 21: 5-167.

_______ (1976). Prehistoric and Protohistoric Remains. *Orissa Review*. Bhubaneshwar: 1-14.

(1909, rpt. 1984). *Imperial Gazetteer of India*. Provincial Series *Bengal* Vol. I. Delhi.

(1908). *The Imperial Gazetteer of India*. Provincial Series *North-West Frontier Province*. Calcutta.

(1908). *The Imperial Gazetteer of India.*Oxford.

(1879). *The Rajputana Gazetteer* Vol. I. Calcutta.

(1948). *The Wealth of India* Raw Materials I. Delhi.

(1950). *The Wealth of India* Raw Materials II. Delhi.

Tindal, A.S. and Brown, G. (1979). The Measurement of the Silver Concentrations of Coins using a Small Neutron Source. *Archaeometry* 21; 27-46.

Traill, G.W. (1878a). Statistical Sketch of Kumaun. *Official Reports on the Province of Kumaun*. Calcutta: 2-52.

______ (1878b). Report on the Bhotea Mahals of Kumaun. *Official Reports on the Province of Kumaun* (J.H. Batten), Calcutta: 62-89.

Tripathi, V. (1973). Introduction of Iron in India. *Radiocarbon and Indian Archaeology* (D.P. Agrawal and A. Ghosh). Bombay: 272-8.

______ (1976). *The Painted Grey Ware An Iron Age Culture of Northern India*. Delhi.

Tripathi, V. (1998). *Archaeometallurgy in India*. Delhi.

Trivedi, R.K. (1965). *Gujarat State Gazetteers Rajkot District*. Ahmedabad.

______ (1969). *Gujarat State Gazetteers Bhavnagar Districti*. Ahmedabad.

Tusa, S. (1979). The Swat Valley in the 2nd and 1st Millennia B.C.: A Question of Marginality, *SAA* 1977 (M. Teddei). Naples: 675-93.

______ (1981). Notes on Some Protohistoric Finds in the Swat Valley (Pakistan). *EW* 31: 99-120.

Varadan, V.K.S. (1976a). *Geology and Mineral Resources of the States of India. Jammu and Kashmir State. G.S.I. Miscellaneous Publication No. 30*. Calcutta.

______ (1976b). *Geology and Mineral Resources of the States of India. Kerala. G.S.I. Miscellaneous Publication No. 30*. Calcutta.

Vardarajan, L. (1983). *India in the 17th Century* (Social, Economic and Political) Memoirs of Francois Martin (1670-1694) Volume I Part-II (1674-1684). New Delhi.

Vemban, N.A. and Yellur, D.D. (1972). Geology of Lead-Zinc Deposit at Khandia, Baroda district, Gujarat. *G.S.I. Miscellaneous Publication No. 16 Base Metals* (Part II). Delhi: 538-52.

Wakankar, V.S. (1967). Kayatha Excavation Number. *The Vikram Journal of Vikram University, Ujjain*. Ujjain.

______ (1968-9). New Light on Central Indian Archaeology Through Kayatha Excavations. *Puratattva* 2: 26-9.

______ (1982). Chalcolithic Malwa. *Indian Archaeology New Perspectives* (R.K. Sharma). Delhi: 224-37.

Walton, H.G. (1910). *British Garhwal A Gazetteer... Agra and Oudh*. Allahabad.

_____ (1911). *Almora: A Gazetteer... Agra and Oudh*. Allahabad.

Watt, G. (1889, rpt. 1972). *A Dictionary of the Economic Products of India*. Vol. II. Delhi.

Wheeler, R.E.M. (1947-8). Brahmagiri and Chandravalli 1947: Megalithic And Other Cultures in the Chitaldrug District, Mysore State. *AI* 4: 180-310.

_____ (1954). *Rome Beyond The Imperial Frontiers*. London.

_____ (1962). *Charsada A Metropolis of the North West Frontier*. London.

Willies, L., Craddock, P.T., Gujar, L.J. and Hegde, K.T.M. (1984). Ancient Lead and Zinc mining in Rajasthan, India. *WA* 16: 222-33.

Wyttenbach, A. and Schubiger, P.A. (1973). Trace Element Content of Roman Lead by Neutron Activation Analysis. *Archaeometry* 15: 199-207.

Young, G.M. (1946). A New Hoard from Taxila (Bhir Mound). *AI* 1: 27-36.

Yule, P. (1984). On the State of Research of Early South Asian Metalwork. *SAA 1981*. Cambridge: 318.

Ziauddin, M. and Narayanaswami, S. (1974). Gold Resources of India. *Bulletins of the Geological Survey of India Series A—Economic Geology* 38. Delhi.

NOTES ON METALLURGICAL EQUILIBRIUM PHASE DIAGRAM

I

Fe-O-C Equilibrium diagram

It is a phase diagram setting out the equilibrium relationship between composition, temperature and structure of iron, oxygen and carbon. This diagram represents that the reduction of Fe_2O_3 to Fe being topo-chemical means it get reduced in the order $Fe_2O_3 \rightarrow Fe_3O_4 \rightarrow FeO \rightarrow Fe$ irrespective of the chemical reaction being carried out by solid carbon (direct reduction reaction) or by CO gas (indirect reduction reaction). The optimum thermochemical conditions required for the reduction process could be determined from the Fe-O-C equilibrium diagram plotted as a function of CO/CO_2 ratio and temperature. From this diagram it is obvious that as the temperature increases, the stability of CO gas also increases with respect to CO_2 and beyond 1000^{0}C only CO gas is found in the gas at the tuyere level. Hence, the rate controlling step for the total combustion of carbon into CO. The CO/CO_2 equilibrium line crosses the FeO-Fe phase boundary at a point

'P' which represents the minimum conditions for the reduction of iron oxide to iron. The reduction kinetics of iron oxide in CO/CO_2 gas mixture, the optimum temperature required for fast reduction is between 900 to 1100⁰C shown as shaded area in the diagram. In this temperature range the CO/CO_2 ratio is 4:1.

II

FeO – Fe_2O_3 binary phase diagram with PO_2 Occurring at various temperatures

This phase diagram was ploted by K. Nagata of Japan in 1998 has measured the temperature as well as partial pressure of oxygen in front of the tuyere in his experimental 'Tatara' furnace and in presence of the slag containing 75.8% FeO, the partial pressure of oxygen has been found to be $1x10^{-11}$ atm. at 1300⁰C and partial pressure of oxygen of $1x10^{-5}$ t at 1100⁰C prevailing 25 cm above the tuyere level.

III

Copper-Nickel complete solid solubility graph

Solid solution is a metallic alloy in which the microstructure consists of similar crystals with no evidence of separate existence of either alloying element, or of intermetallic compounds. Most alloy systems exhibit solid solubility effects over certain range of composition; some such as the Copper-Nickel system of solubility graph exhibit complete solid solubility before 1083⁰C. Solid solutions may be substitutional if the atomic radii of the constituents do not differ by more than about 12%. If there is a big discrepancy in size factor they may form interstitial solutes. They may also be ordered or disordered as shown, a change in the degree of ordering often affecting damping capacity and magnetic properties.

IV

Lead-Tin Equilibrium Diagram

The melting point of lead is 327°C and tin melts at 232°C and their mixture is solder. Thus solder is an alloy of relative low melting point used to make capillary joints between metal components. Basically there is a soft soldering (with tin-lead alloys as seen in their equilibrium diagram) and hard soldering (otherwise known as brazing conducted with brasses etc.). In ancient India soft soldering was conducted with a copper bit, usually containing 2-5% tin to reduce its rate of dissolution in molten solder. In both the cases of soft and hard soldering a flux is required.

V

Guillet diagram

It is for slow-cooled iron-carbon-nickel alloys which showing the influence of carbon and nickel contents on the microstructure of a steel for a given cooling rate.

VI

Copper-Tin Equilibrium diagram

The copper-tin equilibrium diagram confirmed the pronounced separation between liquidus and solidus results in heavy coring of cast alloys, with the result that beta phases forms in practice if the tin content exceeds 7 or 8%. This phase breaks down to gamma at 586°C which in turn undergoes a entectoid transformation to alpha + delta at 520°C. The beta is basically $Cu_{31}Sn_8$, hard and light blue under the microscope. Although the solubility of tin in copper rapidly decreases below 350°C (becoming approx. 1% at room temperature), transformations are very sluggish below 500⁰C and the epsilon phase (Cu_2Sn) only appears after long annealing treatments. Here liquidus is a line of equilibrium or constitution diagram which marks the temperature-composition relationship above which the only stable phases are liquid, while solidus is the locus of points on an equilibrium diagram below which the only stable phases are solid.

VII

Iron-Carbon equilibrium diagram

The iron carbon equilibrium/phase diagram representing the composition of steels, because carbon is the most important alloying element which, with iron makes steel. So-called plain carbon steels are those in which the principal alloying element is non-metallic carbon. Forming Fe_3C (cementite) in the steel, it both strengthens it and renders it amenable to heat treatment. Pure iron can exist in three different phases, labelled alpha, gamma and delta in order of increasing temperature. Other elements normally present are manganese 0.3 – 0.8% (to convert harmful iron sulphide into relatively innocuous mangnaese sulphide and to deoxidise the steel), silicon 0.1 – 0.5% (to deoxidise the steel), and sulphur and phosphorus each 0.03 – 0.05% (residual impurities).

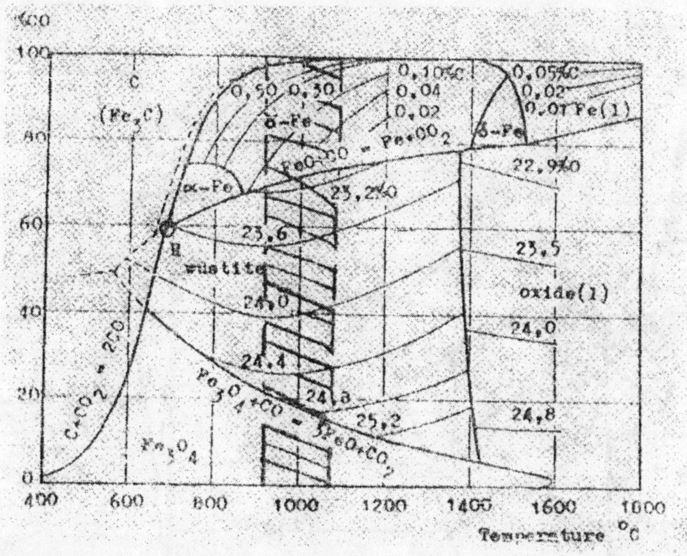

Fe-C-O equilibrium diagram showing the optimum conditions for the bloomery furnace operation, indicated by hatched band

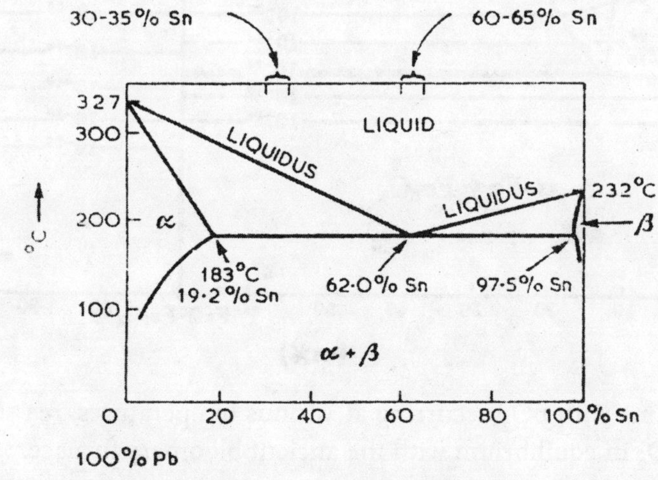

Lead–Tin Equilibrium Diagram.

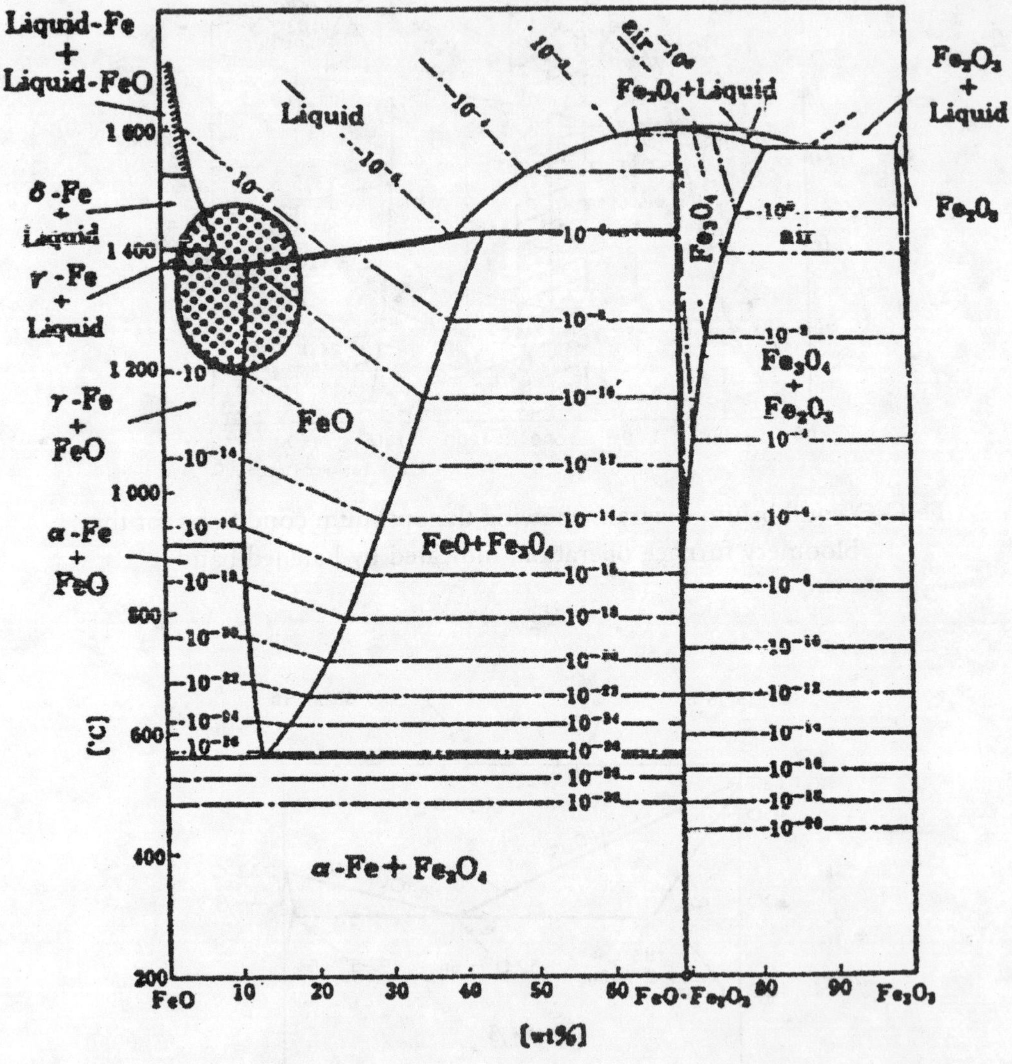

Phase diagrams with pO_2 occurring at various temperatures. rea shows the pO_2 in equilibrium with the ancient bloomery furnace.

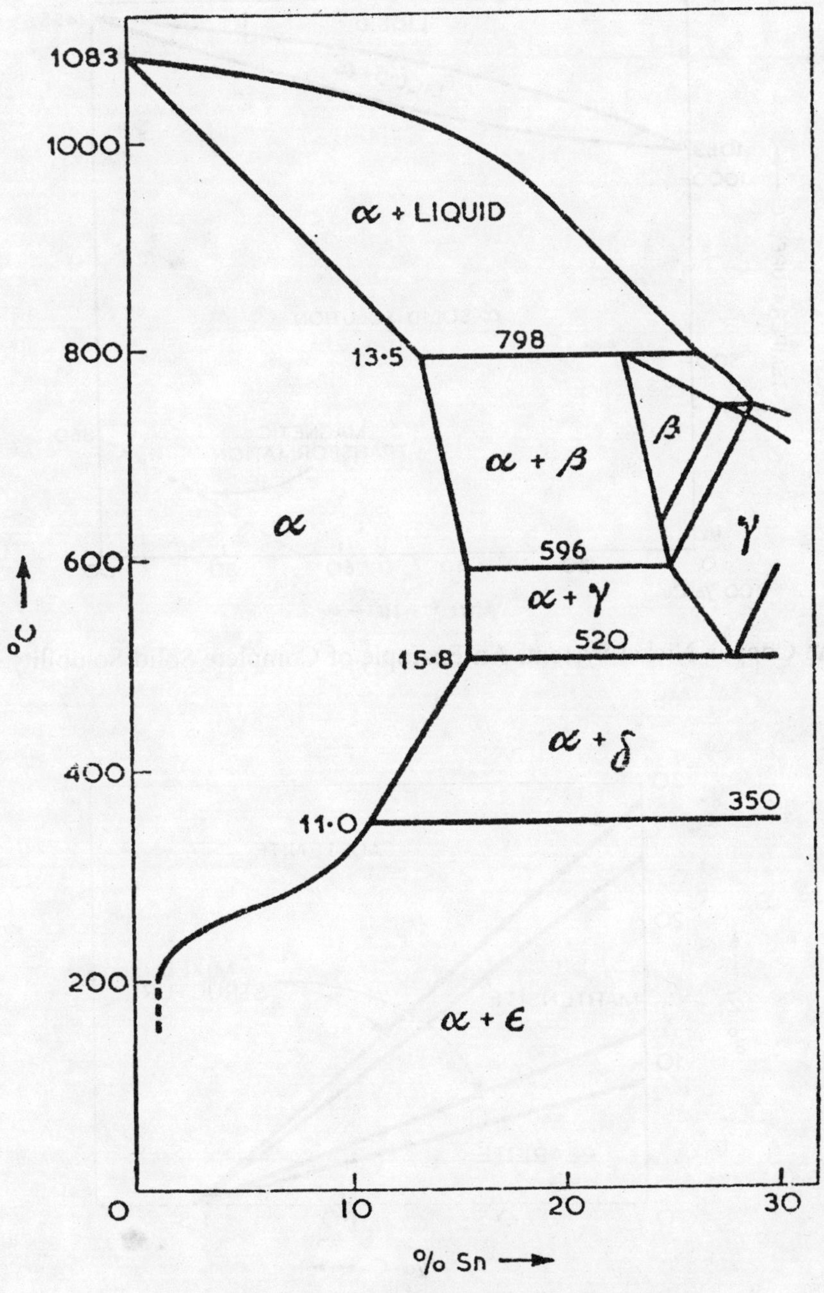

Part of Copper–Tin Equilibrium Diagram.

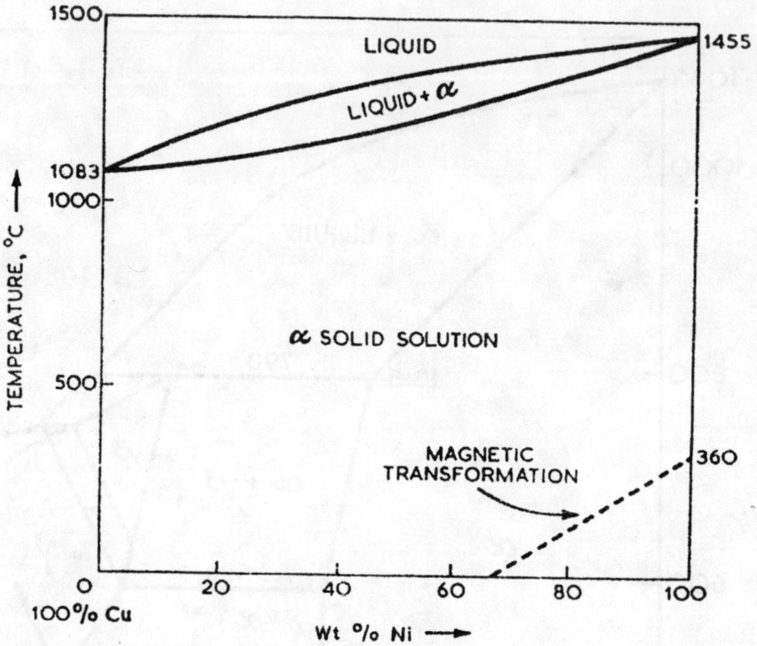

The Cooper-Nickel System, An Example of Complete Solid Solubility.

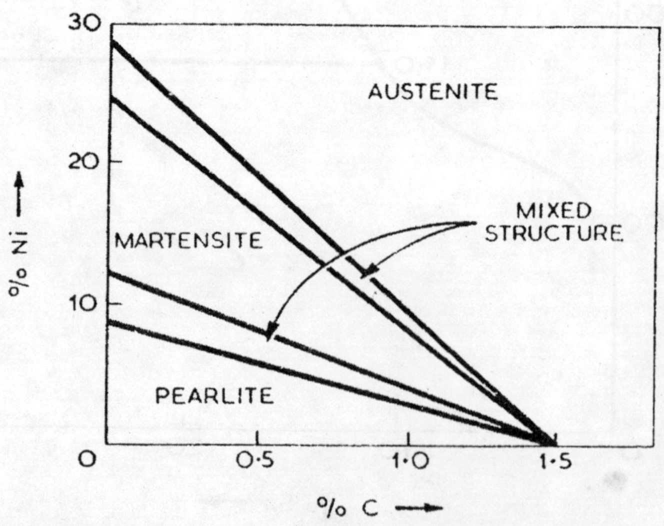

Guillet Diagram for Slow-cooled Iron–Carbon–Nickel Alloys.

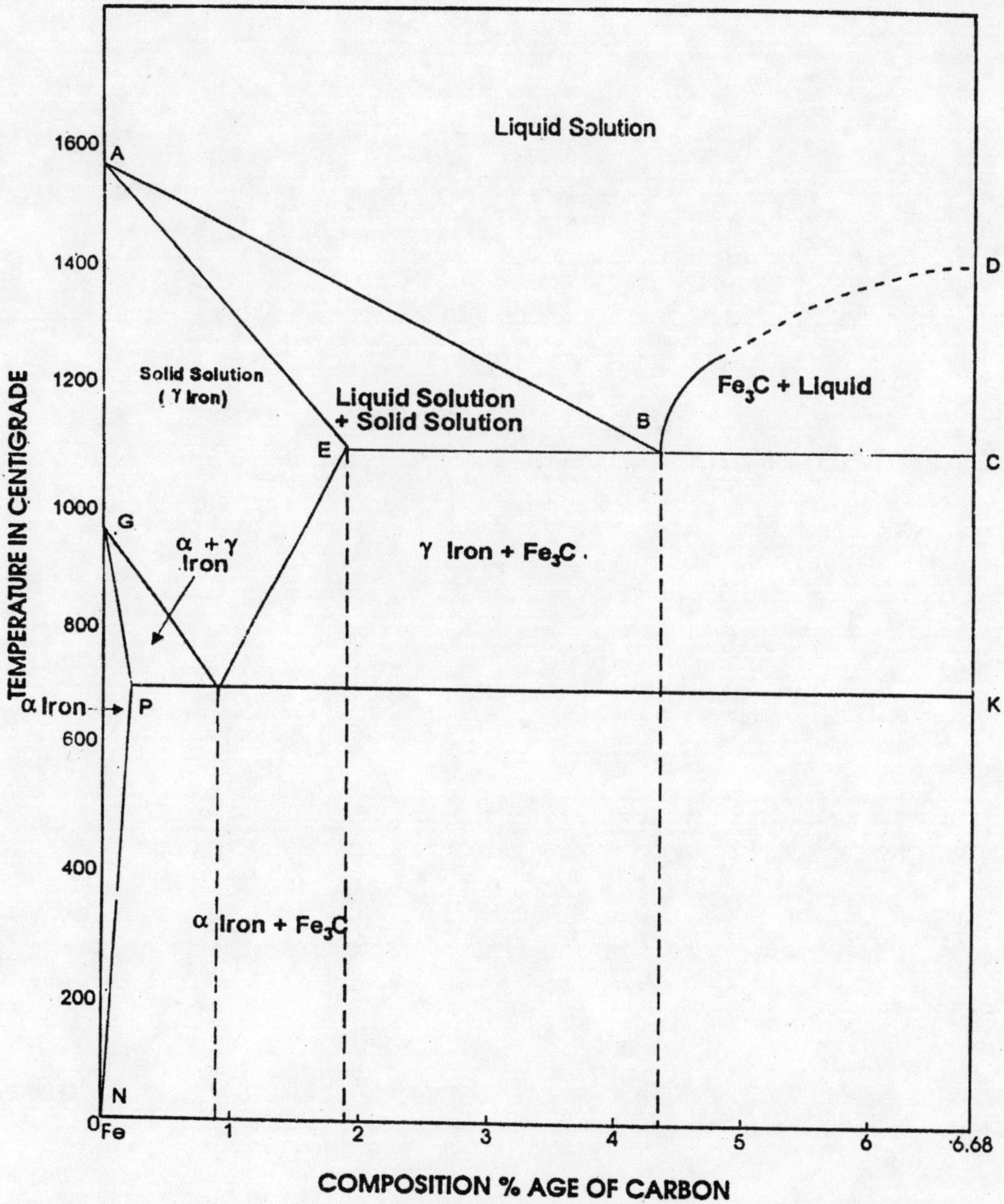

Composition % age of Carbon Graphical Representation of Iron-carbon Alloy of Murai's Tribe (Distt. Bastar)